Analyzing Baseball Data with R

Chapman & Hall/CRC
The R Series

Series Editors

John M. Chambers
Department of Statistics
Stanford University
Stanford, California, USA

Torsten Hothorn
Division of Biostatistics
University of Zurich
Switzerland

Duncan Temple Lang
Department of Statistics
University of California, Davis
Davis, California, USA

Hadley Wickham
Department of Statistics
Rice University
Houston, Texas, USA

Aims and Scope

This book series reflects the recent rapid growth in the development and application of R, the programming language and software environment for statistical computing and graphics. R is now widely used in academic research, education, and industry. It is constantly growing, with new versions of the core software released regularly and more than 4,000 packages available. It is difficult for the documentation to keep pace with the expansion of the software, and this vital book series provides a forum for the publication of books covering many aspects of the development and application of R.

The scope of the series is wide, covering three main threads:
- Applications of R to specific disciplines such as biology, epidemiology, genetics, engineering, finance, and the social sciences.
- Using R for the study of topics of statistical methodology, such as linear and mixed modeling, time series, Bayesian methods, and missing data.
- The development of R, including programming, building packages, and graphics.

The books will appeal to programmers and developers of R software, as well as applied statisticians and data analysts in many fields. The books will feature detailed worked examples and R code fully integrated into the text, ensuring their usefulness to researchers, practitioners and students.

Published Titles

Analyzing Baseball Data with R, *Max Marchi and Jim Albert*

Customer and Business Analytics: Applied Data Mining for Business Decision Making Using R, *Daniel S. Putler and Robert E. Krider*

Dynamic Documents with R and knitr, *Yihui Xie*

Event History Analysis with R, *Göran Broström*

Programming Graphical User Interfaces with R, *Michael F. Lawrence and John Verzani*

R Graphics, Second Edition, *Paul Murrell*

Reproducible Research with R and RStudio, *Christopher Gandrud*

Statistical Computing in C++ and R, *Randall L. Eubank and Ana Kupresanin*

The R Series

Analyzing Baseball Data with R

Max Marchi

Jim Albert

CRC Press
Taylor & Francis Group
Boca Raton London New York

CRC Press is an imprint of the
Taylor & Francis Group an **informa** business

A CHAPMAN & HALL BOOK

CRC Press
Taylor & Francis Group
6000 Broken Sound Parkway NW, Suite 300
Boca Raton, FL 33487-2742

Printed on acid-free paper
Version Date: 20160115

International Standard Book Number-13: 978-1-4665-7022-1 (Paperback)

Visit the Taylor & Francis Web site at
http://www.taylorandfrancis.com

and the CRC Press Web site at
http://www.crcpress.com

Contents

Preface   xv

1 The Baseball Datasets   1

   1.1 Introduction   2
   1.2 The Lahman Database: Season-by-Season Data   2
        1.2.1 Bonds, Aaron, and Ruth home run trajectories   2
        1.2.2 Obtaining the database   4
        1.2.3 The Master table   4
        1.2.4 The Batting table 6
        1.2.5 The Pitching table   8
        1.2.6 The Fielding table 11
        1.2.7 The Teams table   11
        1.2.8 Baseball questions 13
   1.3 Retrosheet Game-by-Game Data   14
        1.3.1 The 1998 McGwire and Sosa home run race   14
        1.3.2 Retrosheet 14
        1.3.3 Game logs   15
        1.3.4 Obtaining the game logs from Retrosheet   16
        1.3.5 Game log example 16
        1.3.6 Baseball questions 16
   1.4 Retrosheet Play-by-Play Data   18
        1.4.1 Event files   18
        1.4.2 Event example   19
        1.4.3 Baseball questions 20
   1.5 Pitch-by-Pitch Data   21
        1.5.1 MLBAM Gameday and PITCHf/x   21
        1.5.2 PITCHf/x Example 22
        1.5.3 Baseball questions 24
   1.6 Summary   25
   1.7 Further Reading   26
   1.8 Exercises   26

2 Introduction to R   29

2.1   Introduction   30
2.2   Installing R and RStudio   30
2.3   Vectors   31
    2.3.1   Career of Warren Spahn   31
    2.3.2   Vectors: defining and calculations   32
    2.3.3   Vector functions   34
    2.3.4   Vector index and logical variables   35
2.4   Objects and Containers in R   36
    2.4.1   Character data and matrices   37
    2.4.2   Factors   38
    2.4.3   Lists   40
2.5   Collection of R Commands   41
    2.5.1   R scripts   41
    2.5.2   R functions   42
2.6   Reading and Writing Data in R   43
    2.6.1   Importing data from a file   43
    2.6.2   Saving datasets   45
2.7   Data Frames   45
    2.7.1   Introduction   45
    2.7.2   Manipulations with data frames   47
    2.7.3   Merging and selecting from data frames   49
2.8   Packages   50
2.9   Splitting, Applying, and Combining Data   50
    2.9.1   Using `sapply`   51
    2.9.2   Using `ddply` in the `plyr` package   52
2.10   Getting Help   54
2.11   Further Reading   55
2.12   Exercises   55

3 Traditional Graphics   59

3.1   Introduction   59
3.2   Factor Variable   60
    3.2.1   A bar graph   60
    3.2.2   Add axes labels and a title   61
    3.2.3   Other graphs of a factor   62
3.3   Saving Graphs   62
3.4   Dot plots   63
3.5   Numeric Variable: Stripchart and Histogram   65
3.6   Two Numeric Variables   67
    3.6.1   Scatterplot   67
    3.6.2   Building a graph, step-by-step   69
3.7   A Numeric Variable and a Factor Variable   73

	3.7.1	Parallel stripcharts	74
	3.7.2	Parallel boxplots	74
3.8	Comparing Ruth, Aaron, Bonds, and A-Rod		76
	3.8.1	Getting the data	76
	3.8.2	Creating the player data frames	78
	3.8.3	Constructing the graph	78
3.9	The 1998 Home Run Race		79
	3.9.1	Getting the data	79
	3.9.2	Extracting the variables	81
	3.9.3	Constructing the graph	82
3.10	Further Reading		82
3.11	Exercises		83

4 The Relation Between Runs and Wins **87**

4.1	Introduction	87
4.2	The Teams Table in Lahman's Database	88
4.3	Linear Regression	89
4.4	The Pythagorean Formula for Winning Percentage	93
4.5	The Exponent in the Pythagorean Formula	95
4.6	Good and Bad Predictions by the Pythagorean Formula	96
4.7	How Many Runs for a Win?	99
4.8	Further Reading	102
4.9	Exercises	102

5 Value of Plays Using Run Expectancy **105**

5.1	The Run Expectancy Matrix		105
5.2	Runs Scored in the Remainder of the Inning		106
5.3	Creating the Matrix		107
5.4	Measuring Success of a Batting Play		110
5.5	Albert Pujols		111
5.6	Opportunity and Success for All Hitters		114
5.7	Position in the Batting Lineup		116
5.8	Run Values of Different Base Hits		119
	5.8.1	Value of a home run	119
	5.8.2	Value of a single	121
5.9	Value of Base Stealing		123
5.10	Further Reading and Software		126
5.11	Exercises		126

6 Advanced Graphics **129**

6.1	Introduction	129
6.2	The `lattice` Package	130

	6.2.1	Introduction	130
	6.2.2	The `verlander` dataset	130
	6.2.3	Basic plotting with `lattice`	132
	6.2.4	Multipanel conditioning	133
	6.2.5	Superposing group elements	134
	6.2.6	Scatterplots and dot plots	135
	6.2.7	The `panel` function	137
	6.2.8	Building a graph, step-by-step	139
6.3	The `ggplot2` Package	144
	6.3.1	Introduction	144
	6.3.2	The `cabrera` dataset	145
	6.3.3	The first layer	146
	6.3.4	Grouping factors	148
	6.3.5	Multipanel conditioning (faceting)	149
	6.3.6	Adding elements	150
	6.3.7	Combining information	151
	6.3.8	Adding a smooth line with error bands	151
	6.3.9	Dealing with cluttered charts	153
	6.3.10	Adding a background image	155
6.4	Further Reading	157
6.5	Exercises	157

7	**Balls and Strikes Effects**	**161**
7.1	Introduction	161
7.2	Hitter's Counts and Pitcher's Counts	162
	7.2.1	Introduction	162
	7.2.2	An example for a single pitcher	162
	7.2.3	Pitch sequences on Retrosheet	165
		7.2.3.1 Functions for string manipulation	165
		7.2.3.2 Finding plate appearances going through a given count	167
	7.2.4	Expected run value by count	169
	7.2.5	The importance of the previous count	170
7.3	Behaviors by Count	173
	7.3.1	Swinging tendencies by count	173
		7.3.1.1 Propensity to swing by location	173
		7.3.1.2 Effect of the ball/strike count	176
	7.3.2	Pitch selection by count	178
	7.3.3	Umpires' behavior by count	181
7.4	Further Reading	184
7.5	Exercises	185

8 Career Trajectories   **187**

8.1 Introduction 187
8.2 Mickey Mantle's Batting Trajectory 188
8.3 Comparing Trajectories 192
    8.3.1 Some preliminary work 192
    8.3.2 Computing career statistics 194
    8.3.3 Computing similarity scores 195
    8.3.4 Defining age, OBP, SLG, and OPS variables 197
    8.3.5 Fitting and plotting trajectories 198
8.4 General Patterns of Peak Ages 202
    8.4.1 Computing all fitted trajectories 202
    8.4.2 Patterns of peak age over time 203
    8.4.3 Peak age and career at-bats 204
8.5 Trajectories and Fielding Position 205
8.6 Further Reading 208
8.7 Exercises 209

9 Simulation   **211**

9.1 Introduction 211
9.2 Simulating a Half Inning 212
    9.2.1 Markov chains 212
    9.2.2 Review of work in runs expectancy 213
    9.2.3 Computing the transition probabilities 215
    9.2.4 Simulating the Markov chain 216
    9.2.5 Beyond runs expectancy 219
    9.2.6 Transition probabilities for individual teams 220
9.3 Simulating a Baseball Season 223
    9.3.1 The Bradley-Terry model 223
    9.3.2 Making up a schedule 224
    9.3.3 Simulating talents and computing win probabilities . . 225
    9.3.4 Simulating the regular season 225
    9.3.5 Simulating the post-season 226
    9.3.6 Function to simulate one season 227
    9.3.7 Simulating many seasons 228
9.4 Further Reading 231
9.5 Exercises 232

10 Exploring Streaky Performances   **237**

10.1 Introduction 237
10.2 The Great Streak 238
    10.2.1 Finding game hitting streaks 238
    10.2.2 Moving batting averages 240

10.3 Streaks in Individual At-Bats   242
     10.3.1 Streaks of hits and outs   242
     10.3.2 Moving batting averages   243
     10.3.3 Finding hitting slumps for all players   243
     10.3.4 Were Suzuki and Ibanez unusually streaky?   246
10.4 Local Patterns of Weighted On-Base Average   249
10.5 Further Reading   255
10.6 Exercises   257

11 Learning About Park Effects by Database Management Tools   **259**

11.1 Introduction   259
11.2 Installing MySQL and Creating a Database   260
11.3 Connecting R to MySQL   262
     11.3.1 Connecting using package `RMySQL`   262
     11.3.2 Connecting using Package `RODBC`   263
11.4 Filling a MySQL Game Log Database from R   264
     11.4.1 From Retrosheet to R   265
     11.4.2 From R to MySQL   265
11.5 Querying Data from R   268
     11.5.1 Introduction   268
     11.5.2 Coors Field and run scoring   271
11.6 Baseball Data as MySQL Dumps   273
     11.6.1 Lahman's database   273
     11.6.2 Retrosheet database   274
     11.6.3 PITCHf/x database   274
11.7 Calculating Basic Park Factors   275
     11.7.1 Loading the data in R   275
     11.7.2 Home run park factor   276
     11.7.3 Assumptions of the proposed approach   277
     11.7.4 Applying park factors   278
11.8 Further Reading   279
11.9 Exercises   279

12 Exploring Fielding Metrics with Contributed R Packages   **283**

12.1 Introduction   283
12.2 A Motivating Example: Comparing Fielding Metrics   284
     12.2.1 Introduction   284
     12.2.2 The fielding metrics   285
     12.2.3 Reading an Excel spreadsheet (`XLConnect`)   286
     12.2.4 Summarizing multiple columns (`doBy`)   287
     12.2.5 Finding the most similar string (`stringdist`)   288
     12.2.6 Applying a function on multiple columns (`plyr`) . . .   291

12.2.7 Weighted correlations (`weights`) 291
12.2.8 Displaying correlation matrices (`ellipse`) 292
12.2.9 Evaluating the fielding metrics (`psych`) 293
12.3 Comparing Two Shortstops 294
12.3.1 Reshaping the data (`reshape2`) 296
12.3.2 Plotting the data (`ggplot2` and `directlabels`) 296
12.4 Further Reading 297
12.5 Exercises 298

A Retrosheet Files Reference **301**

A.1 Downloading Play-by-Play Files 301
A.1.1 Introduction 301
A.1.2 Setup 302
A.1.3 Using a special function for a particular season 302
A.1.4 Reading the files into R 302
A.1.5 The function `parse.retrosheet.pbp` 302
A.2 Retrosheet Event Files: a Short Reference 304
A.2.1 Game and event identifiers 304
A.2.2 The state of the game 305
A.3 Parsing Retrosheet Pitch Sequences 306
A.3.1 Introduction 306
A.3.2 Setup 306
A.3.3 Evaluating every count 307

B Accessing and Using MLBAM Gameday and PITCHf/x Data **311**

B.1 Introduction 311
B.2 Where are the Data Stored? 312
B.3 Suitable Formats for PITCHf/x Data 314
B.3.1 Obtaining data from on-line resources 314
B.3.2 Parsing in R 314
B.3.2.1 A wrapper function 315
B.4 Details on the Data 316
B.4.1 `atbat` attributes 316
B.4.2 `pitch` attributes 317
B.4.3 `hip` attributes (hit locations data) 318
B.5 Special Notes About the Gameday and PITCHf/x Data . . . 319
B.6 Miscellanea 320
B.6.1 Calculating the pitch trajectory 320
B.6.2 An R package for getting and visualizing PITCHf/x data: `pitchRx` 321
B.6.3 Cross-referencing with other data sources 323
B.6.4 Online resources 323

Bibliography   **325**

Index   **329**

Preface

Baseball has always had a fascination with statistics. Schwarz (2005) documents the quantitative measurements of teams and players since the beginning of professional baseball history in the 19th century. Since the foundation of the Society of Baseball Research in 1971, an explosion of new measures have been developed for understanding offensive and defensive contributions of players. One can learn much about the current developments in sabermetrics by viewing articles at websites such as `www.baseballprospectus.com`, `www.hardballtimes.com`, and `www.fangraphs.com`.

The quantity and detail of baseball data has exhibited remarkable growth since the birth of the Internet. The first data were collected for players and teams for individual seasons – this type of data were what would be displayed on the back side of a Topps baseball card. The volunteer-run Project Scoresheet organized the collection of play-by-play game data, and these type of data are currently freely available at the Retrosheet organization at `www.retrosheet.org/`. Since 2006, PITCHf/x data have been measuring the speeds and trajectories of every pitched ball, and newer types of data are collecting the speeds and locations of batted balls and the locations and movements of fielders.

The ready availability of these large baseball datasets has led to challenges for the baseball enthusiast interested in answering baseball questions with these data. It can be problematic to download and organize the data. Standard statistical software packages may be well-suited for working with small datasets of a specific format, but they are less helpful in merging datasets of different types or performing particular types of analyses, say contour graphs of pitch locations, that are helpful for PITCHf/x data.

Fortunately, a new open-source statistical computing environment, R, has experienced increasing popularity among the statistical and computer science community. R is a system for statistical computation and graphics, and it is a computer language designed for typical and possibly specialized statistical and graphical applications. The software is available for unix, Windows, and Macintosh platforms and is available from `www.r-project.org`.

The public availability of baseball data and the open-source R software is an attractive marriage. R provides a large range of tools for importing, arranging, and organizing large datasets. By the use of built-in functions and collections of packages from the R user-community, one can perform various data and graphical analyses, and communicate this work easily to other baseball enthusiasts over the Internet. One of us recently asked a number of MLB team analytics groups about their use of R and here are some responses:

- "We use: R, MySQL / Oracle, perl, php."

- "We do use R extensively, and it is our primary statistical package. The only other major tool we use is probably Excel."

- "We do use R here. It is our primary statistical package for projects that need something more than the statistical functions in Excel."

- "With the occasional exception of Python+NumPy, (R) is the only statistical programming language or package we use."

- "We do use R, its used in conjunction with Excel for analysis."

It is clear that R is a major tool for the analytical work of MLB teams.

The purpose of this book is to introduce R to sabermetricians, baseball enthusiasts, and students interested in exploring baseball data. Chapter 1 provides an overview of the publicly available baseball datasets and Chapter 2 gives a gentle introduction to the type of data structures and exploratory and data management capabilities of R. One of the strongest features of R is its graphics capabilities – Chapter 3 provides an overview of the traditional graphics functions available in the base package and Chapter 6 introduces more sophisticated graphical displays available through the `lattice` and `ggplot2` packages.

The remainder of the book illustrates the use of R in exploring a number of popular topics in sabermetrics. Two fundamental ideas in sabermetrics are the relationship between runs and wins, and the measurement of the value of baseball events by runs. Chapter 4 explores the famous Pythagorean formula derived by Bill James and Chapters 5 and 7 describe the value of plays and pitch sequences using run expectancy. It is fascinating to explore career performance trajectories of ballplayers and Chapter 8 illustrates the use of R to fit quadratic models to player trajectories. Chapter 9 illustrates the use of R simulation functions to simulate a game of baseball by a Markov chain model, and simulate a season of baseball competition. Baseball fans are interested in streaky patterns of performance of teams and players and Chapter 10 explores methods of describing and understanding the significance of streaky patterns of hitting. Given the large size of baseball datasets, it may be more convenient to work with a database and Chapter 11 illustrates the application of several R packages to interface with a MySQL database. Chapter 12 describes the usefulness of several R packages for exploring fielding statistics. The datafiles available through Retrosheet and MLBAM Gameway and PITCHf/x are relatively sophisticated and the appendix material provides detailed descriptions on downloading and reading this data into R.

The reader is encouraged to work on the book datasets and try out the presented R code as the chapters are read. All of the data files and

R code sections used in the book are available at the GitHub repository at `github.com/maxtoki/baseball_R`. In addition, there is a book blog at `baseballwithr.wordpress.com` where the authors will provide advice on using R in sabermetrics research and keep the reader informed on new developments in R software and baseball datasets.

The authors are very grateful for the efforts of our editor, John Kimmel, who played an important role in our collaboration and provided us with timely reviews that led to significant improvements of the manuscript. We wish to thank Anne and Ramona for encouragement and inspiration. Although the two of us live thousands of miles apart, we share a passion both for statistics, baseball, and the knowledge that one can learn about the game through the exploration of data.

1

The Baseball Datasets

CONTENTS

1.1	Introduction	2
1.2	The Lahman Database: Season-by-Season Data	2
	1.2.1 Bonds, Aaron, and Ruth home run trajectories	2
	1.2.2 Obtaining the database	4
	1.2.3 The Master table	4
	1.2.4 The Batting table	6
	1.2.5 The Pitching table	8
	1.2.6 The Fielding table	11
	1.2.7 The Teams table	11
	1.2.8 Baseball questions	13
1.3	Retrosheet Game-by-Game Data	14
	1.3.1 The 1998 McGwire and Sosa home run race	14
	1.3.2 Retrosheet	14
	1.3.3 Game logs	15
	1.3.4 Obtaining the game logs from Retrosheet	16
	1.3.5 Game log example	16
	1.3.6 Baseball questions	16
1.4	Retrosheet Play-by-Play Data	18
	1.4.1 Event files	18
	1.4.2 Event example	19
	1.4.3 Baseball questions	20
1.5	Pitch-by-Pitch Data	21
	1.5.1 MLBAM Gameday and PITCHf/x	21
	1.5.2 PITCHf/x Example	22
	1.5.3 Baseball questions	24
1.6	Summary	25
1.7	Further Reading	26
1.8	Exercises	26

1.1   Introduction

Baseball's marriage with numbers goes back to the origins of the sport. The pioneers of the game in the 1840s had not yet decided the ultimate distance between the pitcher's rubber and home plate, nor the number of balls needed to be awarded a base, when the first box scores and the first stats appeared in newspapers.

This chapter introduces three sources of freely available data, the Lahman database, Retrosheet data, and PITCHf/x data. Baseball records from these sources have a growing level of detail, from seasonal stats available since the 1871 season, to box score data for individual games, to play-by-play accounts covering most games since 1945, to extremely detailed pitch-by-pitch data recorded for nearly all the pitches thrown in MLB parks since 2008. Examples throughout this book will predominately use subsets of data coming from these three sources.

1.2   The Lahman Database: Season-by-Season Data

1.2.1   Bonds, Aaron, and Ruth home run trajectories

In the 2007 baseball season Barry Bonds became the new home run king surpassing Hank Aaron's record of 755 career home runs. Aaron had held the throne since 1974, when he had moved past the legendary Babe Ruth with his 715th home run. Figure 1.1 plots the cumulative home runs of Bonds, Aaron, and Ruth as a function of age. It is clear from the graph that the home run careers of the three sluggers have followed different paths. Aaron was the clear home run leader until age 32 and Aaron and Ruth had similar career home run paths until retirement. Bonds was far behind Aaron and Ruth in career home runs in his 30s, but narrowed the gap and overtook the two sluggers in his 40s.

Babe Ruth began his career as a teenage pitcher for the Boston Red Sox in the so-called Deadball Era when home runs were rare. Ruth's home run impact was not felt until his sixth season, when he began sending the ball out of the park with regularity and outslug nearly every other American League team with 29 home runs. Given his late start, his career line is S-shaped due to his slow start and inevitable decline at the end of his career.

Hank Aaron also made his MLB debut at a very young age and shows a nearly straight line in the graph for the best part of his career. His pattern of hitting home runs was marked by consistency as he hit between 30 and 50 home runs for most seasons of his career. Similar to the Babe, Aaron also

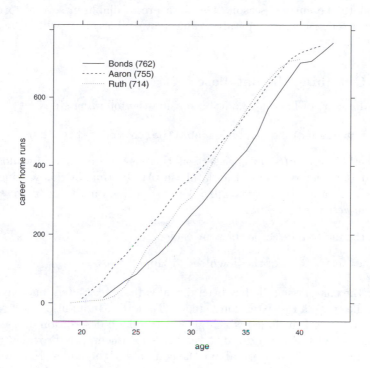

FIGURE 1.1

Career home runs by age for the top three home run hitters in baseball history.

declined in the final years of his career, hitting 20, 12, and 10 home runs from 1974 to 1976.

Barry Bonds had a relatively late major league debut as he did not come to an agreement with the team that first drafted him, and was not in the career home run race until after his 35th birthday. Towards the end of his career, Bonds put together an impressive season home run counts of 49, 73, 46, 45, and 45 home runs, closing in on Babe's 714 mark. Then, after missing most of the 2005 season because of injuries, he completed the chase to the record with two solid seasons (26 and 28 homers) when he was 42 and 43 years old.

To compare sluggers, a researcher needs season-to-season batting data including age and home run counts for Bonds, Aaron, and Ruth. One needs this data for a wide range of seasons, as Ruth's career began in 1914 and Bonds' career ended in 2007.

For many years database journalist and author Sean Lahman has been making available at his website[1] a database (Lahman, 2012) containing pitching, hitting, and fielding statistics for the entire history of professional baseball

---

[1] www.seanlahman.com/

from 1871 to the current season. The data are available in several formats, including a set of comma-separated-value (csv) tables, that will be used in this book.

1.2.2   Obtaining the database

Download a copy of Lahman's database using the following steps.

1. Access `www.seanlahman.com/baseball-archive/statistics/`.

2. Below the *Limited Use License* section, there is the section for downloading the most recent version of the database. At the time of this writing, the section is named *Download 2012 Version*. Click on the *2012 Version - comma-delimited version*.

3. Save the file to a directory of your choice.

4. Extract the content of the downloaded zipped file.

After the compressed file has been download and extracted, one has a total of 24 files with a csv extension listed in Table 1.1. In addition, a "readme 2012.txt" text file is included which gives a thorough description of the Lahman database. One is encouraged to read the documentation provided in the "readme 2012.txt" file to learn about the contents of these files. Here we give a general description of the variables in the data tables most relevant for the studies described in this book.

1.2.3   The Master table

The Master table (`Master.csv` file) is a registry of baseball people. It contains bibliographical information on every player and manager who have appeared at the Major League Baseball level and of all people who have been inducted in the Baseball Hall of Fame.[2] Each line of the `Master.csv` file constitutes a short biography of a person, reporting on dates and places of birth and death, height and weight, throwing hand and batting side, and the dates of the first and last game played.

Players are identified throughout the pitching, batting, and fielding tables in the Lahman's database by an id code, and the Master table is useful for retrieving the name of the player associated with a particular identifier. The table also reports player identification codes of other databases, in particular the ones used by Retrosheet, so one can link players from the Lahman and Retrosheet databases.

For illustration purposes, we display below the header and first line of the

---

[2]Examples of people who never played Major League Baseball but have been inducted in the Hall of Fame (therefore having an entry in the Master table) are baseball pioneer Henry Chadwick and career Negro Leaguer Josh Gibson.

TABLE 1.1

Files in the Lahman's database.

File	Description
AllStarFull.csv	Players' appearances in All-Star Games
Appearances.csv	Seasonal players' appearances by position
AwardsManagers.csv	Recipients of the Manager of the Year Award
AwardsPlayers.csv	Players recipients of the various Awards
AwardsShareManagers.csv	Voting results for the Manager of the Year Award
AwardsSharePlayers.csv	Voting results for the various Awards for players
Batting.csv	Seasonal batting statistics
BattingPost.csv	Seasonal batting statistics for post-season
Fielding.csv	Seasonal fielding statistics
FieldingOF.csv	Seasonal appearances at the three outfield positions
FieldingPost.csv	Seasonal fielding data for post-season
HallOfFame.csv	Voting results for the Hall of Fame
Managers.csv	Seasonal data for managers
ManagersHalf.csv	Seasonal split data for managers
Master.csv	Biographical information for individuals appearing in the database
Pitching.csv	Seasonal pitching statistics
PitchingPost.csv	Seasonal pitching statistics for post-season
Salaries.csv	Seasonal salaries for players
Schools.csv	List of college teams
SchoolsPlayers.csv	Information on schools attended by players
SeriesPost.csv	Outcomes of post-season series
Teams.csv	Seasonal stats for teams
TeamsFranchises.csv	Timelines of Franchises
TeamsHalf.csv	Seasonal split stats for teams

`Master.csv` file which gives information about the first player in the database Hank Aaron. For clarity, we place Aaron's information in a table format in Table 1.2.

```
lahmanID,playerID,managerID,hofID,birthYear,birthMonth,birthDay,
birthCountry,birthState,birthCity,deathYear,deathMonth,deathDay,
deathCountry,deathState,deathCity,nameFirst,nameLast,nameNote,
nameGiven,nameNick,weight,height,bats,throws,debut,finalGame,
college,lahman40ID,lahman45ID,retroID,holtzID,bbrefID
1,aaronha01,,aaronha01h,1934,2,5,USA,AL,Mobile,,,,,,,Hank,Aaron,
,Henry Louis,"Hammer,Hammerin' Hank,Bad Henry",180,72,R,R,
4/13/1954,10/3/1976,,aaronha01,aaronha01,aaroh101,aaronha01,
aaronha01
```

From this information, we learn some details about Aaron's life. Hank Aaron was born February 5, 1934, in Mobile, Alabama, his full name is Henry Louis Aaron, and his nicknames were Hammer, Hammerin' Hank, and Bad Henry. Aaron weighed 180 pounds and was 72 inches tall, he threw and batted right-handed, and he played in the big leagues from 4/13/1954 to 10/3/1976. There is a series of blank columns (those consecutive commas between "Mobile" and "Hank") corresponding to death information, which is obviously unavailable for a living person. Finally there are the various identifying codes for the player. The value of `playerID`, aaronha01, is the identifying code for Hank Aaron in every table in the Lahman's database. The value of the variable `retroID`, aaroh101, is the player id specific to the Retrosheet files to be described in Section 1.3.

1.2.4   The Batting table

The csv file `Batting.csv` file contains all players' batting statistics by season and team from 1871 to the present season. Players in this file are identified with their `playerID`; for example, the season batting statistics of Hank Aaron appear in these table with the identification `playerID = aaronha01`. Each line of the file contains the statistics compiled by a player, during a single season (variable `yearID`), for a particular team (variable `teamID`).

Players who changed teams during a particular season have multiple lines for the season. The `stint` variable indicates the order in which the player moved between teams. For example, Lou Brock, who moved during the 1964 season from the Chicago Cubs to the St. Louis Cardinals, has the following batting lines for the 1964 season.

```
      playerID yearID stint teamID lgID, ...
9567 brocklo01   1964     1    CHN   NL, ...
9568 brocklo01   1964     2    SLN   NL, ...
```

Batting statistics variables are identified by their traditional abbreviations such as AB, R, H, 2B, etc., so the column names of the batting tables should be

TABLE 1.2
First line of the `Master.csv` file.

field name	value
lahmanID	1
playerID	aaronha01
managerID	
hofID	aaronha01h
birthYear	1934
birthMonth	2
birthDay	5
birthCountry	USA
birthState	AL
birthCity	Mobile
deathYear	
deathMonth	
deathDay	
deathCountry	
deathState	
deathCity	
nameFirst	Hank
nameLast	Aaron
nameNote	
nameGiven	Henry Louis
nameNick	Hammer,Hammerin' Hank,Bad Henry
weight	180
height	72
bats	R
throws	R
debut	4/13/1954
finalGame	10/3/1976
college	
lahman40ID	aaronha01
lahman45ID	aaronha01
retroID	aaroh101
holtzID	aaronha01
bbrefID	aaronha01

easily understood by those familiar with baseball box scores. If one has doubts about the meaning of one particular column name, the "readme 2012.txt" file provided with the database gives the variable descriptions.

An excerpt of the `Batting.csv` file for Babe Ruth is conveniently format-ted in Table 1.3. This table shows his batting statistics for his early seasons as a Boston Red Sox pitcher, his years for the Yankees when he became a great home run slugger, and his seasons at the twilight of his career with the Boston Braves.

Only count statistics such as the count of at-bats and count of hits are reported in the batting table. Derived statistics such as a batting average need to be computed from these count statistics. For example, a researcher who wants to know Ruth's batting average for the 1919 season has to calculate it following paragraph 10.21(b) of the *Official Baseball Rules* (Triumph Books, 2012) that instructs to "divide the number of safe hits by the total times at bat." The relevant columns are H and AB, and the desired result is 139 / 432 = .322. Some statistics are not visible for Babe Ruth as they were not recorded in the 1920s. For example, the counts of intentional walks (IBB) are blank for Ruth's seasons, indicating that intentional walks were not recorded for Ruth's seasons.

1.2.5   The Pitching table

The `Pitching.csv` file contains season-by-season pitching data for players. This file contains the traditional count data for pitching such as W (number of wins), L (number of losses), G (games played), BB (number of walks), and SO (number of strikeouts). In addition, this dataset contains several derived statistics such as ERA (earned run average) and BAOpp (opponent's batting average).

Babe Ruth also provides a good illustration of the pitching statistics tables of the Lahman's database since he had a great pitching record before becoming one of the greatest home run hitters in history. Table 1.4 displays statistics from the data file `Pitching.csv` for the seasons where Ruth was a pitcher. We see from the table that Ruth pitched in more than 40 games in 1916 and 1917 (by viewing column G), mostly as a starter (see GS), then appeared on the mound for half that many in the final two seasons for the Red Sox. When he moved to New York he was only an occasional pitcher. Note that Ruth always was a winning pitcher, as his wins (W) outnumbered his losses (L) for all pitching seasons, even when he returned to the pitching mound at the end of his career. He pitched one game both in 1930 and in 1933 (over ten years after he was a dominant pitcher for the Red Sox) and went the full nine innings (see variable CG) in each occasion.

TABLE 1.3

Batting statistics for Babe Ruth, taken from the `Batting.csv` file.

	playerID	yearID	stint	teamID	lgID	G	G_batting	AB	R	H	2B	3B	HR	RBI	SB	CS	BB	SO	IBB	HBP	SH	SF	GIDP	G_old
1	ruthba01	1914	1	BOS	AL	5	5	10	1	2	1	0	0	2	0	0	0	4		0	0	0		5
2	ruthba01	1915	1	BOS	AL	42	42	92	16	29	10	1	4	21	0	0	9	23		0	2	0		42
3	ruthba01	1916	1	BOS	AL	67	67	136	18	37	5	3	3	15	0	0	10	23		0	4	0		67
4	ruthba01	1917	1	BOS	AL	52	52	123	14	40	6	3	2	12	0	0	12	18		0	7	0		52
5	ruthba01	1918	1	BOS	AL	95	95	317	50	95	26	11	11	66	6		58	58		2	3	0		95
6	ruthba01	1919	1	BOS	AL	130	130	432	103	139	34	12	29	114	7		101	58		6	3	0		130
7	ruthba01	1920	1	NYA	AL	142	142	457	158	172	36	9	54	137	14	14	150	80		3	5	0		142
8	ruthba01	1921	1	NYA	AL	152	152	540	177	204	44	16	59	171	17	13	145	81		4	4	0		152
9	ruthba01	1922	1	NYA	AL	110	110	406	94	128	24	8	35	99	2	5	84	80		1	4	0		110
10	ruthba01	1923	1	NYA	AL	152	152	522	151	205	45	13	41	131	17	21	170	93		4	3	0		152
11	ruthba01	1924	1	NYA	AL	153	153	529	143	200	39	7	46	121	9	13	142	81		4	6	0		153
12	ruthba01	1925	1	NYA	AL	98	98	359	61	104	12	2	25	66	2	4	59	68		2	6	0		98
13	ruthba01	1926	1	NYA	AL	152	152	495	139	184	30	5	47	150	11	9	144	76		3	10	0		152
14	ruthba01	1927	1	NYA	AL	151	151	540	158	192	29	8	60	164	7	6	137	89		0	14	0		151
15	ruthba01	1928	1	NYA	AL	154	154	536	163	173	29	8	54	142	4	5	137	87		3	8	0		154
16	ruthba01	1929	1	NYA	AL	135	135	499	121	172	26	6	46	154	5	3	72	60		3	13	0		135
17	ruthba01	1930	1	NYA	AL	145	145	518	150	186	28	9	49	153	10	10	136	61		1	21	0		145
18	ruthba01	1931	1	NYA	AL	145	145	534	149	199	31	3	46	163	5	4	128	51		1	0	0		145
19	ruthba01	1932	1	NYA	AL	133	133	457	120	156	13	5	41	137	2	2	130	62		2	0	0		133
20	ruthba01	1933	1	NYA	AL	137	137	459	97	138	21	3	34	103	4	5	114	90		2	0	0		137
21	ruthba01	1934	1	NYA	AL	125	125	365	78	105	17	4	22	84	1	3	104	63		2	0	0		125
22	ruthba01	1935	1	BSN	NL	28	28	72	13	13	0	0	6	12	0		20	24		0	0	0	2	28

TABLE 1.4

Pitching statistics for Babe Ruth, taken from the `Pitching.csv` file. A few extra columns, not reported here, are available in the `Pitching.csv` file.

	playerID	yearID	stint	teamID	lgID	W	L	G	GS	CG	SHO	SV	IPouts	H	ER	HR	BB	SO	BAOpp	ERA	IBB	WP	HBP	BK	BFP	GF	R
1	ruthba01	1914	1	BOS	AL	2	1	4	3	1	0	0	69	21	10	1	7	3	0.23	3.91		0	0	0	100	0	12
2	ruthba01	1915	1	BOS	AL	18	8	32	28	16	1	0	653	166	59	3	85	112	0.21	2.44		9	6	1	895	3	80
3	ruthba01	1916	1	BOS	AL	23	12	44	41	23	9	1	971	230	63	0	118	170	0.20	1.75		3	8	1	1300	3	83
4	ruthba01	1917	1	BOS	AL	24	13	41	38	35	6	2	979	244	73	2	108	128	0.21	2.01		5	11	0	1313	2	91
5	ruthba01	1918	1	BOS	AL	13	7	20	19	18	1	0	499	125	41	1	49	40	0.21	2.22		3	2	1	660	0	51
6	ruthba01	1919	1	BOS	AL	9	5	17	15	12	0	1	400	148	44	2	58	30	0.29	2.97		5	2	1	591	2	59
7	ruthba01	1920	1	NYA	AL	1	0	1	1	0	0	0	12	3	2	0	2	0	0.20	4.50		0	0	0	17	0	4
8	ruthba01	1921	1	NYA	AL	2	0	2	1	0	0	0	27	14	9	1	9	2	0.35	9.00		0	0	0	49	1	10
9	ruthba01	1930	1	NYA	AL	1	0	1	1	1	0	0	27	11	3	0	2	3	0.30	3.00		0	0	0	39	0	3
10	ruthba01	1933	1	NYA	AL	1	0	1	1	1	0	0	27	12	5	0	3	0	0.30	5.00		0	0	0	42	0	5

1.2.6 The Fielding table

The `Fielding.csv` file contains season-to-season fielding statistics for all players in major league history. For a given player, there will be a separate line for each fielding position. Outfielders positions are grouped together and labeled as `OF` for the older seasons, whereas for the more recent ones they are conveniently distinguished as `LF`, `CF`, `RF`, for left fielders, center fielders, and right fielders, respectively. For a player in a position, the data files give the count of games played (`G`), the count of games started (`GS`), the time played in the field expressed in terms of outs (`InnOuts`), the count of putouts (`PO`), assists (`A`), and errors (`E`).

To illustrate fielding data, Table 1.5 displays Babe Ruth's fielding statistics for his career. Only one line appears for the each of the seasons between 1914 and 1917, as The Babe was exclusively employed as a pitcher. Later, as the Boston Red Sox took advantage of his powerful bat, there are three lines for 1918, one for each defensive position played by Ruth during this season. Suppose one focuses on Ruth's fielding as an outfielder. One raw way of measuring his fielding range, proposed by Bill James in 1977 in his first *Baseball Abstract*, is to sum his putouts (variable `PO`) and assists (variable `A`) and divide the sum by the games played (`G`). The values of this range statistic for the seasons 1918 through 1935 were

```
2.19 2.13 1.99 2.40 2.18 2.69 2.36 2.27 2.14 2.26 2.03 1.84 1.92
1.70 1.71 1.70 1.80 1.54
```

Clearly, Ruth's range as an outfielder deteriorated towards the end of his career.

1.2.7 The Teams table

The `Teams.csv` file contains seasonal data at the team level going back to 1871. A single line in this file includes the team's abbreviation (`teamID`), its final position in the standings (`rank`), its number of wins and losses (`W` and `L`), and whether the team won the World Series (`WSWin`), the League (`LgWin`), the Division (`DivWin`), or reached the post-season via the Wild Card (`WCWin`).

In addition, this file includes cumulative team offensive statistics such as counts of runs scored (`R`), hits (`H`), doubles (`2B`), walks (`BB`), strikeouts (`SO`), stolen bases (`SB`), and sacrifice flies (`SF`). Team defensive statistics include opponents runs scored (`RA`), earned runs allowed (`ER`), complete games (`CG`), shutouts (`SHO`), saves (`SV`), hits allowed (`HA`), home runs allowed (`HRA`), strikeouts by pitchers (`SOA`), and walks by pitchers (`BBA`). Team fielding statistics are included such as the counts of errors (`E`), double plays (`DP`), and the fielding percentage (`FP`). Last, this file includes the total home attendance (`attendance`) and the three-year park factors[3] for batters (`BPF`) and pitchers (`PPF`). Teams are identified, in this and other tables in the database, by a

---

[3]See Chapter 11 for an introduction to park factors.

TABLE 1.5

Fielding statistics for Babe Ruth, taken from the `Fielding.csv` file. Columns featuring statistics relevant only to catchers are not reported.

	playerID	yearID	stint	teamID	lgID	POS	G	GS	InnOuts	PO	A	E	DP
1	ruthba01	1914	1	BOS	AL	P	4			0	7	0	0
2	ruthba01	1915	1	BOS	AL	P	32			17	63	2	3
3	ruthba01	1916	1	BOS	AL	P	44			24	83	3	6
4	ruthba01	1917	1	BOS	AL	P	41			19	101	2	4
5	ruthba01	1918	1	BOS	AL	1B	13			130	6	5	8
6	ruthba01	1918	1	BOS	AL	OF	59			121	8	7	3
7	ruthba01	1918	1	BOS	AL	P	20			19	58	6	5
8	ruthba01	1919	1	BOS	AL	1B	5			35	4	1	4
9	ruthba01	1919	1	BOS	AL	OF	111			222	14	1	6
10	ruthba01	1919	1	BOS	AL	P	17			13	35	2	1
11	ruthba01	1920	1	NYA	AL	1B	2			10	0	1	1
12	ruthba01	1920	1	NYA	AL	OF	141			259	21	19	3
13	ruthba01	1920	1	NYA	AL	P	1			1	0	0	0
14	ruthba01	1921	1	NYA	AL	1B	2			8	0	0	0
15	ruthba01	1921	1	NYA	AL	OF	152			348	17	13	6
16	ruthba01	1921	1	NYA	AL	P	2			1	2	0	0
17	ruthba01	1922	1	NYA	AL	1B	1			0	0	0	0
18	ruthba01	1922	1	NYA	AL	OF	110			226	14	9	3
19	ruthba01	1923	1	NYA	AL	1B	4			41	1	1	2
20	ruthba01	1923	1	NYA	AL	OF	148			378	20	11	2
21	ruthba01	1924	1	NYA	AL	OF	152			340	18	14	4
22	ruthba01	1925	1	NYA	AL	OF	98			207	15	6	3
23	ruthba01	1926	1	NYA	AL	1B	2			10	0	0	2
24	ruthba01	1926	1	NYA	AL	OF	149			308	11	7	5
25	ruthba01	1927	1	NYA	AL	OF	151			328	14	13	4
26	ruthba01	1928	1	NYA	AL	OF	154			304	9	8	0
27	ruthba01	1929	1	NYA	AL	OF	133			240	5	4	2
28	ruthba01	1930	1	NYA	AL	OF	144			266	10	10	0
29	ruthba01	1930	1	NYA	AL	P	1			0	4	0	2
30	ruthba01	1931	1	NYA	AL	1B	1			5	0	0	0
31	ruthba01	1931	1	NYA	AL	OF	142			237	5	7	2
32	ruthba01	1932	1	NYA	AL	1B	1			3	0	0	0
33	ruthba01	1932	1	NYA	AL	OF	128			209	10	9	1
34	ruthba01	1933	1	NYA	AL	1B	1			6	0	1	0
35	ruthba01	1933	1	NYA	AL	OF	132			215	9	7	4
36	ruthba01	1933	1	NYA	AL	P	1			1	1	0	0
37	ruthba01	1934	1	NYA	AL	OF	111			197	3	8	0
38	ruthba01	1935	1	BSN	NL	OF	26			39	1	2	0

three-character code (`teamID`). The column `name` in the `Teams.csv` file helps in recognizing clubs by their full name.

To illustrate the teams dataset, we extract the data for one of the greatest teams in baseball history, the 1927 New York Yankees.

```
yearID lgID teamID franchID divID Rank   G Ghome   W   L DivWin WCWin
  1927   AL    NYA      NYY  <NA>    1 155     77 110 44   <NA>  <NA>
LgWin WSWin   R    AB    H X2B X3B  HR  BB  SO SB CS HBP SF  RA  ER ERA
    Y     Y 975 5347 1644 291 103 158 635 605 90 64  NA NA 599 494 3.2
CG SHO SV IPouts   HA HRA BBA  SOA   E  DP   FP               name
82  11 20   4167 1403  42 409 431 196 123 0.96 New York Yankees
               park attendance BPF PPF teamIDBR teamIDlahman45
820 Yankee Stadium I    1164015  98  94      NYY            NYA
```

We see the 1927 Yankees finished the season with a 110-44 record and won the World Series. The "Bronx Bombers" hit 158 home runs, stole 90 bases, and had a total home attendance of 1,164,015.

1.2.8   Baseball questions

The following questions can be answered with Lahman's database.

Q **What is the average number of home runs per game recorded in each decade? Does the rate of strikeouts show any correlation with the rate of home runs?**

A The number of home runs per game soared from 0.3 in baseball's first two decades to 0.8 in the 1920s. After the 1920s, the home run rate showed a steady increase up to 2.2 per game at the turn of the millennium. The first years of the current decade seems to reflect a decline in home run hitting as the rate has decreased to 1.9 HR per game. Strikeouts have steadily increased over the history of baseball – the number of strikeouts per game was 1 in the 1870s to 5.6 in the 1920s to 14.2 of the 2010s.

Relevant data to obtain this answer is found in the `Teams` *table.*

Q **What effect has the introduction of the Designated Hitter (DH) in the American League had in the difference in run scoring between the American and National Leagues?**

A The DH rule was instituted in 1973 only for the American League. Twice in the previous three years the National League teams had scored half a run more per game than the American League teams. From 1973 till the end of the decade runs scoring was roughly equal. Since then, the American League has maintained an edge of about half a run per game.

Relevant data to obtain this answer is found in the `Teams` *table.*

Q **How does the percentage of games completed by the starting pitcher from 2000 to 2010 compare to the percentage of games 100 years before?**

A From 1900 to 1909 pitchers completed 79% of the games they started, from
  2000 to 2010 it had dropped to 3.5%.

  Data for this answer can be found in the `Pitching` *table.*

1.3   Retrosheet Game-by-Game Data

1.3.1   The 1998 McGwire and Sosa home run race

Another sacred Babe Ruth record was the 60 home runs recorded in the 1927
season. This record was eventually broken in 1961 by Roger Maris, after a
thrilling race with his teammate Mickey Mantle: the "M&M Brothers," as they
were often dubbed, ended the season with 61 and 54 home runs, respectively.
The new home run record lasted another 37 years. In 1998 two other players,
Mark McGwire of the St. Louis Cardinals and Sammy Sosa of the Chicago
Cubs, gave life to a new home run race, that is displayed in Figure 1.2. This
graph shows the cumulative home run count of each player as a function of
the day of the 1998 season.

From the figure, we see that for much of the season, McGwire was the only
man in the chase. Then Sosa caught fire and the two were very close in home
runs starting from mid-August. "Big Mac" first broke the record, hitting his
62nd home run on September 8. Then, on September 25, the two were tied at
66 apiece. Finally, McGwire managed to hit four more in the final days of the
season, while "Slammin' Sammy" remained at 66.

To produce the graph in Figure 1.2 and relive the 1998 season, one needs
data at a game-by-game level.

1.3.2   Retrosheet

Retrosheet is a volunteer organization, founded in 1989 by University of
Delaware professor David Smith, that aims to collect play-by-play accounts of
every game played in Major League Baseball history. Through the labor of love
of many volunteers who have unearthed old newspaper accounts, scanned mi-
crofilms, and manually entered data into computers, the Retrosheet website[4]
contains game-by-game summaries going back to the dawn of Major League
Baseball in the nineteenth century. The Retrosheet site also has play-by-play
data of most of the games played since the 1945 season and continues to add
games for previous seasons. This data is introduced in Section 1.4.

---

[4]www.retrosheet.org

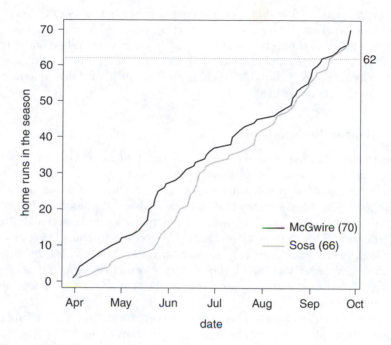

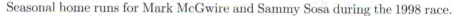

FIGURE 1.2
Seasonal home runs for Mark McGwire and Sammy Sosa during the 1998 race.

1.3.3 Game logs

Retrosheet provides individual game data going back to 1871. A game log has details regarding when the game was played, how many spectators attended, the teams and the ballpark, and the score (both the final score and the inning by inning runs scored). In addition, the game log file include teams offensive and defensive statistics, starting players, managers, and umpire crews. There are missing observations for some game log variables for earlier baseball seasons.

Retrosheet provides a comprehensive *Guide to Retrosheet Game Logs*[5] document which gives details of all 161 fields compiled for each game. Readers are encouraged to peruse the guide to fully understand the contents of the files. Details on the relevant data fields will be described when they are used in later chapters.

---

[5]www.retrosheet.org/gamelogs/glfields.txt

1.3.4 Obtaining the game logs from Retrosheet

Game log files can be found at www.retrosheet.org/gamelogs/index.html. A zip file is provided for each season, starting from 1871, and can be downloaded in a folder of choice by clicking on the relevant year. When one extracts the zip file, one obtains a plain text file (.txt extension) where fields are separated by commas. Appendix A provides an R script file for downloading and parsing all the game log files.

1.3.5 Game log example

On September 9, 1995, Cal Ripken Jr. of the Baltimore Orioles surpassed the seemingly unbeatable consecutive games record of 2130 belonging to the late Lou Gehrig. One can learn more about this historic game by exploring the game log files for the 1995 season. Table 1.6 contains a subset of the copious information available for this particular game between Baltimore and California. This data is taken from a single line in the gl1995.txt file available at www.retrosheet.org/gamelogs/index.html. This table displays team statistics[6] as well as the players' identities and fielding positions for the home team; similar statistics and player information are available for the visitor team.

What does one learn from this game log information displayed in Table 1.6? This game took place on a Wednesday night in front of 46,272 people in Baltimore (the hometeam = BAL indicates the Orioles were the home team). The game lasted over three and a half hours (duration = 215 minutes), thanks in part to the standing ovation Ripken got at the end of the fifth inning, when the game became official. (The standing ovation information is not available in this file.) Baltimore defeated California 4-2; since we observe homehr = 4, we observe that all of Baltimore's runs this game were due to four home runs with the bases empty. The Orioles infield in this game included Rafael Palmeiro at first base, Chris Hoiles at second base, Ripken at shortstop, and Jeff Huson at third base.

1.3.6 Baseball questions

Here are some typical questions one can answer with the Retrosheet game logs files.

Q **In which months are home runs more likely to occur? What about ballparks?**

A Since 1980, July has been the month with the most home runs per game (1.97), while September has had the lowest frequency (1.84). In the same

---

[6]Some other team statistics, such as Stolen Bases and Caught Stealing, omitted in Table 1.6, are reported in Game log files.

TABLE 1.6

Excerpt of information available in the `Game Logs`. Sample from the Cal Ripken's Iron Man game (Sept. 9, 1995).

date	19950906
dayofweek	Wed
visitorteam	CAL
hometeam	BAL
visitorrunsscored	2
homerunsscore	4
daynight	N
parkid	BAL12
attendance	46272
duration	215
visitorlinescore	100000010
homelinescore	10020010x
homeab	34
homeh	9
homehr	4
homerbi	4
homebb	1
homek	8
homegdp	0
homelob	7
homepo	27
homea	8
homee	0
umpirehname	Larry Barnett
umpire1bname	Greg Kosc
umpire2bname	Dan Morrison
umpire3bname	Al Clark
visitormanagername	Marcel Lachemann
homemanagername	Phil Regan
homestartingpitchername	Mike Mussina
homebatting1name	Brady Anderson
homebatting1position	8
homebatting2name	Manny Alexander
homebatting2position	4
homebatting3name	Rafael Palmeiro
homebatting3position	3
homebatting4name	Bobby Bonilla
homebatting4position	9
homebatting5name	Cal Ripken
homebatting5position	6
homebatting6name	Harold Baines
homebatting6position	10
homebatting7name	Chris Hoiles
homebatting7position	2
homebatting8name	Jeff Huson
homebatting8position	5
homebatting9name	Mark Smith
homebatting9position	7

time frame, 2.71 home runs per game have been hit in Coors Field (home of the Colorado Rockies), and 1.14 in the Astrodome (the former home of the Houston Astros).

Q **Do runs happen more frequently when some umpires are behind the plate? What is the difference between the most pitcher-friendly and the most hitter-friendly umpires?**

A Among umpires with more than 400 games called since 1980, teams scored the highest number of runs (10.0 per game combined) when Chuck Meriwether was behind the plate and the lowest (7.8) when Doug Harvey was in charge.

Q **How many extra people attend ballgames during the weekend? What's the average attendance by day of the week?**

A Close to 33,000 people attend games played on Saturdays (data from 1980 to 2011) and 31,000 on Sundays. The average goes down to 29,000 on Fridays, 25,000 on Thursdays and Mondays, and 24,000 on Tuesdays and Wednesdays.

1.4   Retrosheet Play-by-Play Data

1.4.1   Event files

Retrosheet has collected data to an even finer detail for most games played since 1945. For those seasons, play-by-play accounts are available at www.retrosheet.org/game.htm. These "event files" (as these play-by-play files are named) contain information on every single event happening on the field during a game. For each play, information is reported on the situation (inning, team batting, number of outs, presence of runners on base), the players on the field, the sequence of pitches thrown, and details on the play itself. For example, the file indicates whether a hit occurred and if a ball in play is a ground ball, the file gives the defender that fielded the ball.

Event files come in a format expressly devised for them. The process of rendering the files in a format suitable for use in R (or other statistical programs) is not straightforward, thus we will provide a friendly version of the event files for the seasons used in the book. Retrosheet gives detailed instruction on how to use the files[7] and a step-by-step guide,[8] plus the software to parse the files.[9] In Appendix A, R code is presented that implements the full process of downloading, extracting, and parsing data.

---

[7]How to use Our Event Files: www.retrosheet.org/datause.txt
[8]Step-by-Step Example: www.retrosheet.org/stepex.txt
[9]Software Tools: www.retrosheet.org/tools.htm

1.4.2 Event example

As a historical game was used for the purpose of showing the contents of Retrosheet game logs, a famous fielding play is used to illustrate the Retrosheet event files. This play is represented as a single line in an event file shown in Table 1.7.

The play took place in a game played in Oakland on October 13, 2001, as can be inferred from the value of the GAME_ID variable. This game was Game 3 of the American League Division Series featuring the hometown Athletics against the New York Yankees (AWAY_TEAM_ID = NYA). The play occurred in the seventh inning with the home team batting (variables INN_CT and BAT_ID_ID). There were two outs (variable OUTS_CT) and the A's were leading 1-0 (variables AWAY_SCORE_CT and HOME_SCORE_CT). Right-handed Mike Mussina (variables PIT_ID and PIT_HAND_CD) was on the mound for the Yankees, facing left-handed batter Terrence Long (variables BAT_ID and BAT_HAND_CD) with Jeremy Giambi standing on first base (variable BASE1_RUN_ID). The BAT_FLD_CD = 7 and BAT_LINEUP_ID = 7 fields inform us that Giambi's defensive position was left field (position 7 corresponding to left field) and he was batting 7th in the lineup. The variables POS2_FLD_ID through POS9_FLD_ID report the full defensive lineup for the Yankees.

The seemingly unintelligible characters appearing in the PITCH_SEQ_TX and EVENT_TX variables depict what happened during that particular at bat. From looking at the pitch sequence variable PITCH_SEQ_TX, one sees that Mussina quickly went ahead in the count as Long let a strike go by and swung and missed another pitch (CS). Then Mussina followed with consecutive balls (BB) and Long battled with a foul ball (F) before putting the ball in play (X). The variable EVENT_TX gives the results of the play. Long's hit resulted in a double, collected by the Yankees right fielder (D9 in the Event Text) in short right (9S). The runner on first was thrown out on his way to home (1XH) by a throw from right fielder Shane Spencer, relayed by shortstop Derek Jeter to catcher Jorge Posada (962).

Once the event files are properly processed, many more fields are available than the ones presented in Table 1.7. However these additional fields are, for the most part, derived from what is in the table. For example, one additional field indicates whether the at-bat resulted in a base hit, one field will identify the fielder who collected the ball, and four fields will indicate where each runner (and the batter) stood at the end of the play – all of this can be inferred by the EVENT_TX field.

This play-by-play information is available for most games going back to 1945, thus it is possible to recreate what happened on the field in the past half-century. For this particular play, the Retrosheet event files cannot tell us all of the interesting details. Derek Jeter came out of nowhere to cut off

TABLE 1.7
Excerpt of information available in Retrosheet `Event` files. Sample from Jeter's "Flip Play" (Oct. 13, 2001)

1.	GAME_ID	OAK200110130
2.	YEAR_ID	2001
3.	AWAY_TEAM_ID	NYA
4.	INN_CT	7
5.	BAT_HOME_ID	1
6.	OUTS_CT	2
7.	BALLS_CT	2
8.	STRIKES_CT	2
9.	PITCH_SEQ_TX	CSBBFX
10.	AWAY_SCORE_CT	1
11.	HOME_SCORE_CT	0
12.	BAT_ID	longt002
13.	BAT_HAND_CD	L
14.	PIT_ID	mussm001
15.	PIT_HAND_CD	R
16.	POS2_FLD_ID	posaj001
17.	POS3_FLD_ID	martt002
18.	POS4_FLD_ID	soria001
19.	POS5_FLD_ID	bross001
20.	POS6_FLD_ID	jeted001
21.	POS7_FLD_ID	knobc001
22.	POS8_FLD_ID	willb002
23.	POS9_FLD_ID	spens001
24.	BASE1_RUN_ID	giamj002
25.	BASE2_RUN_ID	NA
26.	BASE3_RUN_ID	NA
27.	EVENT_TX	D9/9S.1XH(962)
28.	BAT_FLD_CD	7
29.	BAT_LINEUP_ID	7

Spencer's throw and flipped it backhand to Posada in time to nail Giambi at home, on what has become known as "The Flip Play".[10]

1.4.3   Baseball questions

Below are some questions that can be explored with the Retrosheet event files. These specific questions are about how batters perform in particular situations in the pitch count and with runners on base.

Q **During the McGwire/Sosa home run race, which player was more successful at hitting homers with men on base?**

---

[10]In 2002, *Baseball Weekly* recognized "The Flip Play" as one of the ten most amazing fielding plays of all time.

A Mark McGwire hit 37 home runs in 313 plate appearances with runners on base, Sammy Sosa 29 in 367. Once walks (both intentional and unintentional) and hit by pitches are removed, the number of opportunities become 223 for McGwire and 317 for Sosa.

Q **How many intentional walks in unusual situations (e.g., empty bases or bases loaded) was Barry Bonds issued in his 73 HR campaign?**

A During his record 2001 season, Barry Bonds was passed intentionally only 35 times. Of those free passes one came with a runner on first and two with runners on first and second. When he was awarded 120 intentional walks in 2004, 19 came with nobody on, 11 with a runner on first, and 3 with runners on first and third. He was once walked intentionally with the bases loaded in 1998.

Q **What is the Major league batting average when the ball/strike count is 0-2? What about on 2-0?**

A In 2011, hitters compiled a .253 batting average on plate appearances where they fell behind 0-2. Conversely they hit .479 after going ahead 2-0.

1.5   Pitch-by-Pitch Data

1.5.1   MLBAM Gameday and PITCHf/x

After having hit a combined 59 home runs in four seasons, Blue Jays right-fielder Jose Bautista emerged in 2010 as an elite slugger by blasting a league leading 54 balls out of the park. Figure 1.3 shows the location and the type of the 54 pitches Jose sent into the stands.

Since 2005 baseball fans have had the opportunity to follow, pitch-by-pitch, the games played by their favorite team on the Web thanks to the MLBAM[11] Gameday application featured on the MLB.com website. For a couple of years fans would only know the outcome of each pitch (whether it was a ball, a called strike, a swinging strike, and so on). Starting from an October 2006 game played at the Metrodome in Minneapolis, a wealth of detail began to appear for each pitch tracked on Gameday. Data on the release point, the pitch speed, and its full trajectory, have been available for about one-third of the games played in 2007. Starting from the 2008 season, nearly every MLB pitch flight has been recorded by the PITCHf/x system.

PITCHf/x is a product by Sportvision, a company which produces broadcast effects for sports, such as the first-down virtual line for football and the

---

[11]Major League Baseball Advanced Media

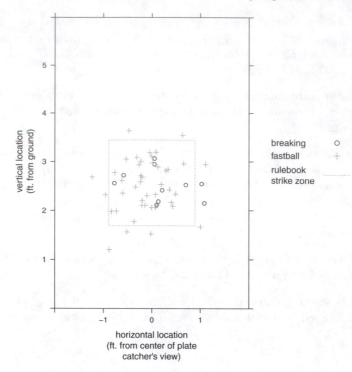

FIGURE 1.3
Pitch type and location for Jose Bautista's 54 home runs of the 2010 season.

FoxTrax hockey puck. Two cameras installed in each MLB park record the flight of the baseball between the pitcher's mound and home plate, and advanced software calculates the position, the velocity and the acceleration of the ball, giving sufficient information to estimate the full trajectory of the ball in its mound-to-plate trip. Raw PITCHf/x data can be accessed by anyone from the MLB.com website, however its format (XML) might not be easy to manage for the average reader. The data used for the examples in this book are available in a format suitable for quick use inside R. In Appendix B we will show how the XML package allows to manage data in XML format in R and point to some free online resources that can be used to download PITCHf/x data.

1.5.2 PITCHf/x Example

On April 21, 2012, Phil Humber became the 21st pitcher in Major League Baseball history to throw a perfect game by retiring all the 27 batters he faced. PITCHf/x captured his final pitch (like it has done for nearly every other pitch thrown in MLB ballparks since 2008), providing the data shown in Table 1.8. The outcome of the pitch (variable `des`) is recorded by a stringer, while most of the remaining information is either captured by the Sportvision system or calculated from the captured data.

Each pitch is assigned an identifier (`sv_id`), that is actually a time stamp: Humber's final pitch was recorded on April 21, 2012, at 15:25:37. The key information Sportvision obtains through its camera system is recorded in lines 11 through 19 of the table. Those nine parameters give the position (variables `x0`, `y0`, `z0`), velocity (variables `vx0`, `vy0`, `vz0`), and acceleration (variables `ax`, `ay`, `az`) components of the pitch at release point. With these nine parameters the full trajectory of the pitch from release to home plate can be estimated. (In fact, Sportvision actually estimates the parameters somewhere in the middle of the ball's flight, then derives the parameters at release point.)

While the nine parameters just mentioned are sufficient for learning about the trajectory of the pitch, they are difficult to understand by casual fans who follow the game on MLBAM Gameday.[12] Other more descriptive quantitities are calculated starting from those nine parameters. The one measure familiar to baseball fans is the pitch speed at release, which for Humber's final pitch is calculated at 85.3 mph (variable `start_speed`). PITCHf/x also provides the speed of the ball as it crosses the plate, 79.1 mph in this case (variable `end_speed`). Another two important values are the variables `px` and `pz`; they represent the horizontal and vertical location of the pitches, respectively, and can be combined with the batter's strike zone upper and lower limits (`sz_top` and `sz_bot`) to infer whether the pitch crossed the strike zone.

Let's focus on the location of this particular pitch. The horizontal reference point is the middle of the plate, with positive values indicating pitches passing on the right side of it from the umpire's viewpoint. In this case the ball crossed the plate 2.21 feet on the right of its midpoint. Since the plate is 17 inches wide, it was way out of the strike zone. The pitch was also too low to be a strike, as the vertical point at which crossed the plate is listed at 1.17 feet, while the hitter's lower limit of the strike zone is 1.74.[13] Luckily for Humber, since otherwise a walk would have ruined the perfect game, the home plate umpire controversially declared that Brendan Ryan had swung the bat for strike three.

Other interesting quantities about a pitch are available with PITCHf/x, including the horizontal and vertical movement (variables `pfx_x` and `pfx_z`) of the pitch trajectory, the spin direction, and its rate (variables `spin_dir` and `spin_rate`).[14] MLBAM has devised a complex algorithm which processes

---

[12]Philip Humber's perfect game can be relived, pitch by pitch, at `mlb.mlb.com/mlb/gameday/index.jsp?gid=2012_04_21_chamlb_seamlb_1&mode=gameday`.

[13]The batter's strike zone boundaries are recorded by the human stringer at the beginning of the at-bat, and thus are less precise than the pitch location coordinates recorded by the advanced system.

[14]Detailed explanations for the PITCHf/x fields have been provided by Mike Fast at `fastballs.wordpress.com/2007/08/02/glossary-of-the-gameday-pitch-fields/`. Prof. Alan Nathan provides a collection of PITCHf/x references at `webusers.npl.illinois.edu/~a-nathan/pob/pitchtracker.html`.

TABLE 1.8

Excerpt of information available from PITCHf/x. Sample from the final pitch
of Phil Humber's perfect game (Apr. 21, 2012)

1. des	Swinging Strike (Blocked)
2. sv_id	120421_152537
3. start_speed	85.3
4. end_speed	79.1
5. sz_top	3.73
6. sz_bot	1.74
7. pfx_x	0.31
8. pfx_z	1.81
9. px	2.211
10. pz	1.17
11. x0	-1.58
12. y0	50.0
13. z0	5.746
14. vx0	9.228
15. vy0	-124.71
16. vz0	-5.311
17. ax	0.483
18. ay	25.576
19. az	-29.254
20. break_y	23.8
21. break_angle	-4.1
22. break_length	7.8
22. pitch_type	SL
23. spin_dir	170.609
24. spin_rate	344.307

the information captured by Sportvision and marks the pitch with a label
familiar to baseball fans. In this case the algorithm recognizes the pitch as a
slider (variable `pitch_type`).

1.5.3   Baseball questions

Below are questions you can answer with PITCHf/x data. The data can be
used to address specific questions about pitch type, speed of the pitch, and
play outcomes on specific pitches.

Q **Who are the hitters who see the lowest and the highest percentage of fastballs?**

A From 2008 to 2011 pitchers have thrown fastballs 35% of the time when
Ryan Howard was at the plate, 56% of the time when facing David Eck-
stein.

Q **Who is the fastest pitcher in baseball currently?**

A Nine of the fastest ten pitches recorded by PITCHf/x from 2008 to 2011 have been thrown by Aroldis Chapman, the highest figure being a 105.1 mph pitch thrown on September 24, 2010, in San Diego. Neftali Feliz is the other pitcher making the top ten list with a 103.4 fastball delivered in Kansas City on September 1, 2010.

Q **What are the chances of a successful steal when the pitcher throws a fastball compared to when a curve is delivered?**

A From 2008 to 2010 baserunners were successful 73% of the times at stealing second base on a fastball. The success rate increases to 85% when the pitch is a curveball.

1.6 Summary

When choosing among the three main sources of baseball data (Lahman, Retrosheet, and PITCHf/x), one always has to consider the trade-off between the level of detail and the seasons covered by the source. With Lahman's database, for example, one can explore the evolution of the game since its beginnings back into the nineteenth century. However only the basic season count statistics are available from this source. For example, simple information such as Babe Ruth's batting splits by pitcher's handedness cannot be retrieved from Lahman's files.

Retrosheet is steadily adding past seasons to its play-by-play database, allowing researchers to perform studies to validate or reject common beliefs about players of the past decades. During the years, for example, analysis of play-by-play data has confirmed the huge defensive value of players like Brooks Robinson and Mark Belanger, and has substantiated the greatness of Roberto Clemente's throwing arm.

PITCHf/x has been available only since 2008 and, contrary to Retrosheet, there is no way to compile data for games of the past. This means we will never be able to compare the velocity of Aroldis Chapman's fastball to that of Nolan Ryan or Bob Feller. However, studies performed since its inception have provided an enhanced understanding of the game, enabling researchers to explore issues like pitch sequencing, batter discipline, pitcher fatigue, and the catcher's ability to block bad pitches.

1.7   Further Reading

Schwarz (2005) provides a detailed history of baseball statistics. Adler (2006) explains how to obtain baseball data from several sources, including Lahman's database, Retrosheet, and MLBAM Gameday and how to analyze them using diverse tools, from Microsoft Excel to R, MySQL and PERL. Fast (2009) introduces the PITCHf/x system to the uninitiated. The *PITCHf/x, HITf/x, FIELDf/x* section of The Physics of Baseball website (Nathan, 2012) features material on the subject of pitch tracking data.

1.8   Exercises

1. (**Which Datafile?**)

   This chapter has given an overview of the Lahman database, the Retrosheet game logs, the Retrosheet play-by-play files, and the PITCHf/x database. Describe the relevant data among these four databases that can be used to answer the following baseball questions.

   (a) How has the rate of walks (per team for nine innings) changed over the history of baseball?

   (b) What fraction of baseball games in 1968 were shutouts? Compare this fraction with the fraction of shutouts in the 2012 baseball season.

   (c) What percentage of first pitches are strikes? If the count is 2-0, what fraction of the pitches are strikes?

   (d) Is it easier to steal second base or third base? (Compare the fraction of successful steals of second base with the fraction of successful steals of third base.)

2. (**Lahman Pitching Data**)

   From the pitching data file from the Lahman database, the following information is collected about Bob Gibson's famous 1968 season.

   ```
   playerID yearID stint teamID lgID  W L  G GS CG SHO SV IPouts
   gibsobo01  1968    1    SLN   NL 22 9 34 34 28  13  0    914
    H ER HR BB  SO BAOpp  ERA IBB WP HBP BK  BFP GF  R SH SF GIDP
   198 38 11 62 268  0.18 1.12   6  4   7  0 1161  0 49 NA NA   NA
   ```

   (a) Gibson started 34 games for the Cardinals in 1968. What fraction of these games were completed by Gibson?

   (b) What was Gibson's ratio of strikeouts to walks this season?

    (c) One can compute Gibson's innings pitched by dividing `IPouts` by three. How many innings did Gibson pitch this season?

    (d) A modern measure of pitching effectiveness is WHIP, the average number of hits and walks allowed per inning. What was Gibson's WHIP for the 1968 season?

3. (**Retrosheet Game Log**)

Jim Bunning pitched a perfect game on Father's Day on June 21, 1964. Some details about this particular game can be found from the Retrosheet game logs.

```
    date  day visitorteam visitorleague visitorgame hometeam
19640621  Sun         PHI            NL          60      NYN
homeleague homegame visitorscore homescore lengthgame daynight
      NL        67            6         0         54        D
completioninfo forfeitinfo protestinfo parkid attendance timegame
          NA          NA          NA   NYC17          0      139
visitorlinescore homelinescore visitorab visitorh visitor2b
       110004000     000000000        32        8         2
visitor3b visitorhr visitorrbi visitorsh visitorsf visitorhbp
        0         1          6         2         0          0
visitorbb visitoribb visitork visitorsb visitorcs visitorgdp
        4          0        6         0         1          0
visitorci visitorlob
        0         5
```

    (a) What was the time in hours and minutes of this particular game?

    (b) Why is the attendance value in this record equal to zero?

    (c) How many extra base hits did the Phillies have in this game? (We know that the Mets had no extra base hits this game.)

    (d) What was the Phillies' on-base percentage in this game?

4. (**Retrosheet Play-by-Play Record**)

One of the famous plays in Philadelphia Phillies baseball history is second-baseman Mickey Morandini's unassisted triple play against the Pirates on September 20, 1992.[15] The following records from the Retrosheet play-by-play database describe this half-inning. The variables indicate the half-inning (variables `INN_CT` and `HOME_ID`), the current score (variables `AWAY_SCORE_CT` and `HOME_SCORE_CT`), the identities of the pitcher and batter (variables `BAT_ID` and `PIT_ID`), the pitch sequence (variable `PITCH_SEQ`), the play event description (variable `EVENT_TEX`), and the runners on base (variables `BASE1_RUN` and `BASE2_ID`).

---

[15]This play is described in detail at `phillysportshistory.com/2011/09/20/`.

```
INN_CT BAT_HOME_ID AWAY_SCORE_CT HOME_SCORE_CT    BAT_ID    PIT_ID
    6           1             1             1 vansa001  schic002
    6           1             1             1 bondb001  schic002
    6           1             1             1 kingj001  schic002
PITCH_SEQ_TX EVENT_CD            EVENT_TX BASE1_RUN_ID BASE2_RUN_ID
       CBBBX       20            S9/L9M
       C1BX        20         S/G56.1-2     vansa001
       BLLBBX       2 4(B)4(2)4(1)/LTP/L4M  bondb001     vansa001
```

Based on the records, write a short paragraph that describes the play-by-play events of this particular inning.

5. (**PITCHf/x Record of Several Pitches**)

R. A. Dickey is one of the current pitchers who predominantly throws a knuckleball. The following gives some PITCHf/x variables for the first knuckleball and the first fastball that Dickey threw for a game against the Kansas City Royals on April 13, 2013.

```
start_speed end_speed pfx_x pfx_z      px    pz sz_bot sz_top
         73      66.3 -0.64 -7.58 -0.047 2.475    1.5   3.35


start_speed end_speed pfx_x pfx_z     px    pz sz_bot sz_top
       81.2      75.4 -4.99 -7.67 -1.99 2.963    1.5   3.43
```

Describe the differences between the knuckleball and the fastball in terms of pitch speed, movement (horizontal and vertical directions), and location in the strike zone. Based on this data, why is a knuckleball so difficult for a batter to make contact?

2

Introduction to R

CONTENTS

2.1	Introduction	30
2.2	Installing R and RStudio	30
2.3	Vectors	31
	2.3.1 Career of Warren Spahn	31
	2.3.2 Vectors: defining and calculations	32
	2.3.3 Vector functions	34
	2.3.4 Vector index and logical variables	35
2.4	Objects and Containers in R	36
	2.4.1 Character data and matrices	37
	2.4.2 Factors	38
	2.4.3 Lists	40
2.5	Collection of R Commands	41
	2.5.1 R scripts	41
	2.5.2 R functions	42
2.6	Reading and Writing Data in R	43
	2.6.1 Importing data from a file	43
	2.6.2 Saving datasets	45
2.7	Data Frames	45
	2.7.1 Introduction	45
	2.7.2 Manipulations with data frames	47
	2.7.3 Merging and selecting from data frames	49
2.8	Packages	50
2.9	Splitting, Applying, and Combining Data	50
	2.9.1 Using `sapply`	51
	2.9.2 Using `ddply` in the `plyr` package	52
2.10	Getting Help	54
2.11	Further Reading	55
2.12	Exercises	55

2.1   Introduction

In this chapter, we provide a general introduction to the R statistical system. We describe the process of installing R and the program RStudio that provides an attractive interface to the R system. Pitching data from the legend Warren Spahn is used to motivate manipulations with vectors, a basic data structure. We describe different data types such as characters, factors, and lists, and different "containers" for holding these different data types. The process of executing collections of R commands by means of scripts and functions is discussed, and methods for importing and exporting datasets from R are described. A fundamental data structure in R is a data frame and we introduce defining a data frame, performing manipulations, merging data frames, and performing operations on a data frame split by values of a variable. We conclude the chapter by describing how to install and load R packages and how one gets help using resources from the R system and the RStudio interface.

2.2   Installing R and RStudio

The R system (R Development Core Team (2013)) is available for download from The Comprehensive R Archive Network (CRAN) at `www.r-project.org`. R is available for Linux, Windows, and Macintosh systems; all of the commands described in this book will work in any of these environments.

One can use R through the standard graphical user interface by launching the R application. Recently, several new developmental environments have been created for R, and we will demonstrate the RStudio environment (RStudio (2013)) available from `rstudio.com`. In installation, one first installs R and then the RStudio application, and then R is opened by launching the RStudio application.

The RStudio opening screen is displayed in Figure 2.1. The screen is divided into four windows. One can type commands directly and see output in the lower-left Console window. Moving clockwise, the top-left window is a blank file where one can write and execute R scripts or groups of instructions. The top-right window shows names of objects such as vectors and data frames created in an R session. By clicking on the History tab, one can see a record of all commands entered during the current R session. Last, any plots are displayed in the lower-right window. By clicking on the Files tab, one can see a list of files stored in the current working directory. (This is the file directory where R will expect to read files, and where any output, such as data files and graphs, will be stored.) The Packages tab lists all of the R packages currently installed in the system and the Help tab will display documentation for R functions and datasets.

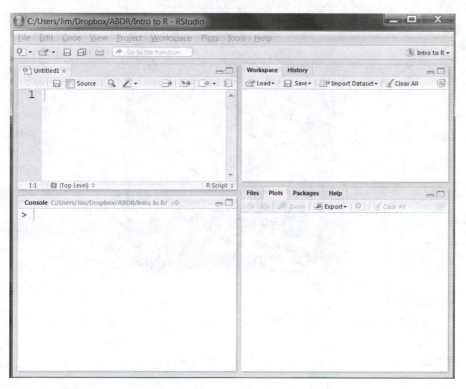

FIGURE 2.1
The opening screen of the RStudio interface to R.

2.3   Vectors

2.3.1   Career of Warren Spahn

One of the authors collected the 1965 Warren Spahn baseball card pictured in Figure 2.2. The back of the card, shown in Figure 2.3, displays many of the standard pitching statistics for the seasons preceding Spahn's final 1965 season. We use data from Spahn's season statistics to illustrate some basic components of the R system.

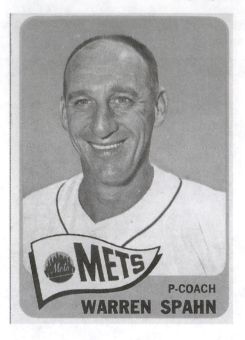

FIGURE 2.2
Front of the 1965 Warren Spahn Topps card.

2.3.2 Vectors: defining and calculations

A basic structure in R is a *vector*, a sequence of values of a given type, such as numeric or character. A basic way of creating a vector is by means of the c (combine) function. To illustrate, suppose we are interested in exploring the games won and lost by Spahn for the seasons after the war when he played for the Boston Braves. We create two vectors by use of the c function; the games won are stored in the vector W and the games lost are stored in the vector L. The symbol "<-" is the assignment character in R; the "=" symbol can also be used for assignment. These lines can be directly typed into the Console window. (By the way, R is case sensitive, so R will distinguish the vector L from the vector l.)

```
W <- c(8, 21, 15, 21, 21, 22, 14)
L <- c(5, 10, 12, 14, 17, 14, 19)
```

One attractive feature of R is its ability to do element-by-element calculations with vectors. Suppose we wish to compute the winning percentage for Spahn for these seven seasons. We want to compute the fraction of winning games and multiply this fraction by 100 to convert it to a percentage. We create a new vector named Win.Pct by use of the basic multiplication (*) and division (/) operators:

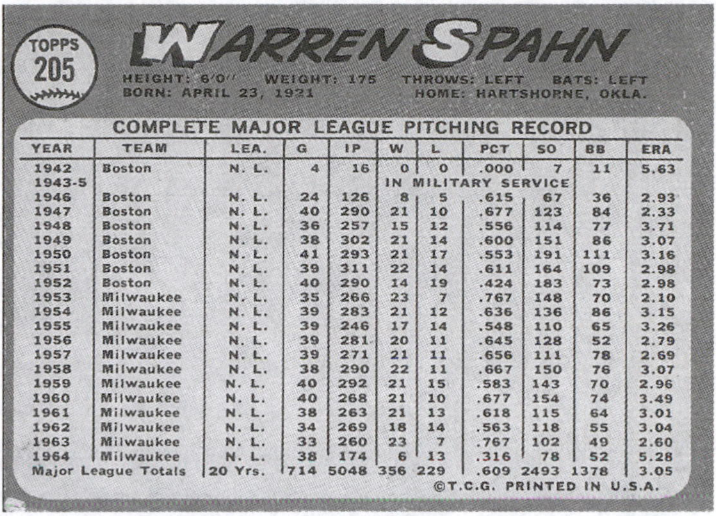

TOPPS 205

WARREN SPAHN

HEIGHT: 6'0" WEIGHT: 175 THROWS: LEFT BATS: LEFT

BORN: APRIL 23, 1921 HOME: HARTSHORNE, OKLA.

COMPLETE MAJOR LEAGUE PITCHING RECORD

YEAR	TEAM	LEA.	G	IP	W	L	PCT	SO	BB	ERA
1942	Boston	N. L.	4	16	0	0	.000	7	11	5.63
1943-5					IN MILITARY SERVICE					
1946	Boston	N. L.	24	126	8	5	.615	67	36	2.93
1947	Boston	N. L.	40	290	21	10	.677	123	84	2.33
1948	Boston	N. L.	36	257	15	12	.556	114	77	3.71
1949	Boston	N. L.	38	302	21	14	.600	151	86	3.07
1950	Boston	N. L.	41	293	21	17	.553	191	111	3.16
1951	Boston	N. L.	39	311	22	14	.611	164	109	2.98
1952	Boston	N. L.	40	290	14	19	.424	183	73	2.98
1953	Milwaukee	N. L.	35	266	23	7	.767	148	70	2.10
1954	Milwaukee	N. L.	39	283	21	12	.636	136	86	3.15
1955	Milwaukee	N. L.	39	246	17	14	.548	110	65	3.26
1956	Milwaukee	N. L.	39	281	20	11	.645	128	52	2.79
1957	Milwaukee	N. L.	39	271	21	11	.656	111	78	2.69
1958	Milwaukee	N. L.	38	290	22	11	.667	150	76	3.07
1959	Milwaukee	N. L.	40	292	21	15	.583	143	70	2.96
1960	Milwaukee	N. L.	40	268	21	10	.677	154	74	3.49
1961	Milwaukee	N. L.	38	263	21	13	.618	115	64	3.01
1962	Milwaukee	N. L.	34	269	18	14	.563	118	55	3.04
1963	Milwaukee	N. L.	33	260	23	7	.767	102	49	2.60
1964	Milwaukee	N. L.	38	174	6	13	.316	78	52	5.28
Major League Totals		20 Yrs.	714	5048	356	229	.609	2493	1378	3.05

©T.C.G. PRINTED IN U.S.A.

FIGURE 2.3
Back of the 1965 Warren Spahn Topps card.

```
Win.Pct <- 100 * W / (W + L)
```

We can display these winning percentages by simply typing the variable name:

```
Win.Pct
[1] 61.53846 67.74194 55.55556 60.00000 55.26316 61.11111 42.42424
```

A convenient way of creating patterned data is by use of the function `seq`. We use this function to generate the season years from 1946 to 1952 and store the output to the variable `Year`.[1]

```
Year <- seq(1946, 1952)
Year
[1] 1946 1947 1948 1949 1950 1951 1952
```

For a sequence of integer values, the colon notation will also work:

```
Year <- 1946 : 1952
```

Suppose we wish to calculate Spahn's age for these seasons. Spahn was born in April 1921 and we can compute his age by subtracting 1921 from each season value – the resulting vector is stored in the variable `Age`.

```
Age <- Year - 1921
```

We construct a simple scatterplot of Spahn winning percentages (vertical) against his age (horizontal) by use of the `plot` function (see Figure 2.4).

[1] The function `seq(a, b, s)` will generate a vector of values from `a` to `b` in steps of `s`.

```
plot(Age, Win.Pct)
```

We see that Spahn was pretty successful for most of his Boston seasons – his winning percentage exceeded 55% for six of his seven seasons.

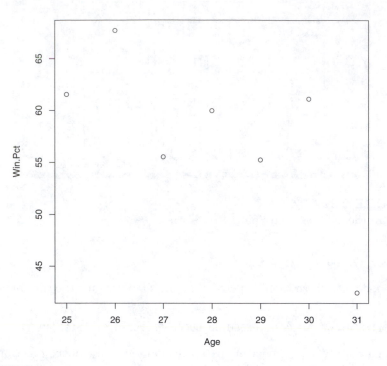

FIGURE 2.4
Scatterplot of the winning percentage against age for Warren Spahn's seasons playing for the Boston Braves.

2.3.3   Vector functions

There are many built-in R functions for vectors including `mean` (arithmetic average), `sd` (standard deviation), `length` (number of vector entries), `sum` (sum of values), `max` (maximum value), `sort`, and `order`. To illustrate a function, one can use the `mean` function to find the average winning percentage of Spahn during this seven-season period.

```
mean(Win.Pct)
[1] 57.66207
```

It is actually more common to compute a pitcher's career winning percentage by dividing his win total by the number of wins and losses. One can compute this career winning percentage by means of the following R expression.

```
100 * sum(W) / (sum(W) + sum(L))
[1] 57.277
```

One can sort the win numbers from low to high by the `sort` function:

```
sort(W)
[1]  8 14 15 21 21 21 22
```

The `cumsum` function is useful for displaying cumulative totals of a vector

```
cumsum(W)
[1]   8  29  44  65  86 108 122
```

We see from the output that Spahn won 8 games in the first season, 29 games in the first two seasons, and so on. The `summary` function applied on the winning percentages displays several summary statistics of the vector values such as the extremes (low and high values), the quartiles (first and third), the median, and the mean.

```
summary(Win.Pct)
   Min. 1st Qu.  Median    Mean 3rd Qu.    Max.
  42.42   55.41   60.00   57.66   61.32   67.74
```

This output tells us that his median winning percentage was 60, his mean percentage was 57.66, and the entire group of winning percentages ranged from 42.42 to 67.74.

2.3.4 Vector index and logical variables

To extract portions of vectors, the square bracket is used. For example, the expression

```
W[c(1, 2, 5)]
[1]  8 21 21
```

will extract the first, second, and fifth entries of the vector W. The first four values of the vector can be extracted by typing

```
W[1 : 4]
[1]  8 21 15 21
```

By use of a minus index, we remove entries from a vector. For example, if we wish to remove the first and sixth entries of W, we would type

```
W[-c(1, 6)]
[1] 21 15 21 21 14
```

A logical variable is created in R by the use of a vector together with the operations >, <, == (logical equal), and != (logical not equals). For example, suppose we are interested in the values in the winning percentage vector `Win.Pct` that exceed 60%.

```
Win.Pct > 60
[1]  TRUE  TRUE FALSE FALSE FALSE  TRUE FALSE
```

The result of this calculation is a logical vector – the output indicates that Spahn had a winning percentage exceeding 60% for the first, second, and sixth seasons (TRUE), and not exceeding 60% for the remaining seasons (FALSE). Were there any seasons where Spahn won more than 20 games and his winning percentage exceeded 60%? We use the logical & (AND) operator to find the years where W > 20 and Win.Pct > 60.

```
(W > 20) & (Win.Pct > 60)
[1] FALSE  TRUE FALSE FALSE FALSE  TRUE FALSE
```

The output indicates that both conditions were true for the second and sixth seasons.

By using logical variables and the square bracket notion, we can find subsets of vectors satisfying different conditions. During this period, when did Spahn have his highest winning percentage? We use

```
Win.Pct == max(Win.Pct)
[1] FALSE  TRUE FALSE FALSE FALSE FALSE FALSE
```

to create a logical vector which is true when this condition is satisfied. (Note the use of the double equals sign notion to indicate logical equality.) Then we select the corresponding year by indexing Year by this logical vector.

```
Year[Win.Pct == max(Win.Pct)]
[1] 1947
```

We see that the highest winning percentage occurred in 1947 during this period.

What seasons did the number of decisions (wins plus losses) exceed 30? We first create a logical vector based on W + L > 30, and then choose the seasons using this logical vector.

```
Year[W + L > 30]
[1] 1947 1949 1950 1951 1952
```

We see that the number of decisions exceeded 30 for the five seasons 1947, 1949, 1950, 1951, and 1952.

2.4   Objects and Containers in R

The individual data components such as the years 1947, 1949, and 1950 are called *objects*. These objects can be of different *types* such as numeric, logical, character, and integer. We have already worked with objects of types numeric and logical in the previous section. We store a number of objects into a *container*. A vector is a simple type of container where we place a number of objects of the same type, say objects that are all numeric or all logical. Here we illustrate some of the different object types and containers that we find useful in working with baseball data.

2.4.1 Character data and matrices

String variables such as the names of teams and players are stored as *characters* which are represented by letters and numbers enclosed by double quotes. As a simple example, suppose we wish to explore information about the World Series in the years 2003 through 2012. We create three character vectors NL, AL, and Winner containing abbreviations for the National League winner, the American League winner, and the league of the team that won the World Series. Note that we represent each character value by a string of letters enclosed by double quotes. We also define two numeric vectors: N.Games contains the number of games of each series, and Year gives the corresponding seasons.

```
NL <- c("FLA", "STL", "HOU", "STL", "COL",
        "PHI", "PHI", "SFG", "STL", "SFG")
AL <- c("NYY", "BOS", "CHW", "DET", "BOS",
        "TBR", "NYY", "TEX", "TEX", "DET")
Winner <- c("NL", "AL", "AL", "NL", "NL",
            "NL", "AL", "NL", "NL", "NL")
N.Games <- c(6, 4, 4, 5, 4, 5, 6, 5, 7, 4)
Year <- 2003 : 2012
```

There are other ways to store objects besides vectors. For example, suppose we wish to display the World Series contestants in a tabular format. A *matrix* is a rectangular grid of objects of the same type. A matrix can be created by the matrix function; the arguments are the objects to be put in the matrix and the number of rows and the number of columns. By default the objects are placed in the matrix by columns. Suppose we want to create a matrix with 10 rows and 2 columns with the National League contestants in the first column and the American League contestants in the second column. We combine the two team vectors into one vector by the c function and apply the matrix function, storing the result in variable results.

```
results <- matrix(c(NL, AL), 10, 2)
results
      [,1]  [,2]
 [1,] "FLA" "NYY"
 [2,] "STL" "BOS"
 [3,] "HOU" "CHW"
 [4,] "STL" "DET"
 [5,] "COL" "BOS"
 [6,] "PHI" "TBR"
 [7,] "PHI" "NYY"
```

```
 [8,]  "SFG" "TEX"
 [9,]  "STL" "TEX"
[10,]  "SFG" "DET"
```

One can use the `dimnames` function to add descriptive labels to the rows and columns of the matrix. We label the rows by the seasons using the first index of `dimnames(results)` (note the use of the double square brackets) and label the columns with `"NL Team"` and `"AL Team"`. The matrix is now in a more readable format.

```
dimnames(results)[[1]] <- Year
dimnames(results)[[2]] <- c("NL Team", "AL Team")
results
     NL Team AL Team
2003 "FLA"   "NYY"
2004 "STL"   "BOS"
2005 "HOU"   "CHW"
2006 "STL"   "DET"
2007 "COL"   "BOS"
2008 "PHI"   "TBR"
2009 "PHI"   "NYY"
2010 "SFG"   "TEX"
2011 "STL"   "TEX"
2012 "SFG"   "DET"
```

There are a number of R functions available for exploring character data. The `table` function will construct a frequency table for a vector of character data. For example, to learn about the number of wins by each league in the 10 World Series, we use `table` with the variable Winner

```
table(Winner)
Winner
AL NL
 3  7
```

It is interesting that the National League won 7 of these 10 World Series. If we `barplot` the result from `table`

```
barplot(table(Winner))
```

we obtain a bar graph of the frequencies (see Figure 2.5).

2.4.2   Factors

A *factor* is a special way of representing character data. To motivate the consideration of factors, suppose we construct a frequency table of the National League representatives to the World Series in the character vector `NL`.

```
table(NL)
NL
COL FLA HOU PHI SFG STL
  1   1   1   2   2   3
```

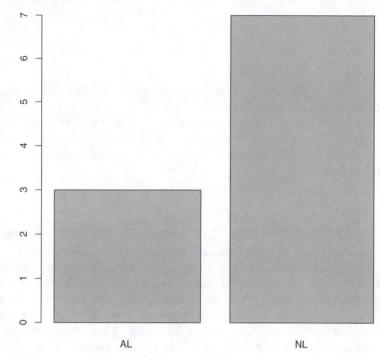

FIGURE 2.5

Bar graph of the number of wins of the American League and National League teams in the World Series between 2003 and 2012.

Note that R will organize the teams alphabetically (from `COL` to `STL`) in the frequency table. It may be preferable to organize the teams by the division (East, Central, and West). We can change the organization of the team labels by converting this character type to a factor.

We make this conversion by means of the `factor` function. The basic arguments are the vector of data to be converted and a vector `levels` that gives the ordered values of the variable. Here we list the values ordered by the East, Central, and West divisions. The result is stored in the factor variable `NL2`.

```
NL2 <- factor(NL, levels=c("FLA", "PHI", "HOU", "STL", "COL", "SFG"))
```

One can understand how factor variables are stored by using the `str` function to examine the structure of the variable `NL2`.

```
str(NL2)
 Factor w/ 6 levels "FLA","PHI","HOU",..: 1 4 3 4 5 2 2 6 4 6
```

We see that a factor variable is actually encoded by integers (1, 4, 3, ...) where the levels are the team names. If we `table` this new factor variable

```
table(NL2)
NL2
FLA PHI HOU STL COL SFG
  1   2   1   3   1   2
```

we obtain the same frequencies as before, but the teams are now listed in the order specified in the `factor` function.

Generally, we will see that R will automatically convert character-type data to factors. Many R functions require the use of factors, and the use of factors gives one finer control on how character labels are displayed in output and graphs.

2.4.3   Lists

All of the containers we have described such as vectors and matrices require that data values have the same type. For example, vectors contain all numeric data or all character data; one cannot mix numeric and character data in a single vector. A *list* is a convenient way of storing data of different types. To illustrate, suppose we wish to collect the league that won the World Series (a character type), the number of games played (a numeric type), and a short description (a character type) into a single variable. Using the `list` function, we create a new list `World.Series` with components `Winner`, `Number.Games`, and `Seasons`.

```
World.Series <- list(Winner=Winner, Number.Games=N.Games,
                     Seasons="2003 to 2012")
```

Once a list such as `World.Series` is defined, there are different ways of accessing the different components. If we wish to display the number of games played `Number.Games`, we can use the list variable name together with with the $ symbol and the component name.

```
World.Series$Number.Games
 [1] 6 4 4 5 4 5 6 5 7 4
```

Or we can use the double square brackets to display the second component of the list.

```
World.Series[[2]]
 [1] 6 4 4 5 4 5 6 5 7 4
```

As an alternative, we can use the single square brackets with the name of the component in quotes.

```
World.Series["Number.Games"]
$Number.Games
 [1] 6 4 4 5 4 5 6 5 7 4
```

Note that the first two options return vectors and the third option returns a list with a single component `Number.Games`.

Many R functions return a list of data of different types, so it is important how to access the list components. Also we will see that lists provide a convenient way of collecting information of different types (character, numeric, logical, factors) about teams and players.

2.5 Collection of R Commands

2.5.1 R scripts

The R expressions described in the previous sections can be typed directly in the Console window and any output will be directly displayed in that window. Alternatively, the R expressions can be stored in a text file, called an R script, and executed as a group of commands.

Continuing our previous example, suppose we wish to run the following R commands. We are entering the length and the winners of ten World Series, tabulating and graphing the winner, and comparing the lengths of the series won by the National and American Leagues. The by function is used to find summaries of the numeric variable N.Games for each level of a categorical variable Winner.

```
N.Games <- c(6, 4, 4, 5, 4, 5, 6, 5, 7, 4)
Winner <- c("NL", "AL", "AL", "NL", "NL",
            "NL", "AL", "NL", "NL", "NL")
table(Winner)
barplot(table(Winner))
by(N.Games, Winner, summary)
```

A convenient way to run R scripts is through the text window in the upper-left window of the RStudio environment. The R commands are typed in this window and the script is executed by selecting these lines and pressing Control-Enter (in a Windows operating system) or Command-Enter (in a Macintosh operating system). The screenshot in Figure 2.6 shows the result of executing this R script. The R output is displayed in the lower-left Command window. In the Workspace window (upper-right), we see that two vectors N.Games and Winner have been created. In the Plots window (lower-right), we see the bar graph as a result of the barplot function.

Another way of running an R script is by saving the commands in a file, and then "sourcing" this file within R. Suppose that a file with the above commands has been saved in the file "World.Series.R" in the current working directory. (Section 2.6.1 explains how one changes the working directory.) Then one can execute this file by typing in the Console window:

```
source("World.Series.R", echo=TRUE)
```

The echo=TRUE argument is used so that the R output is displayed in the Console window.

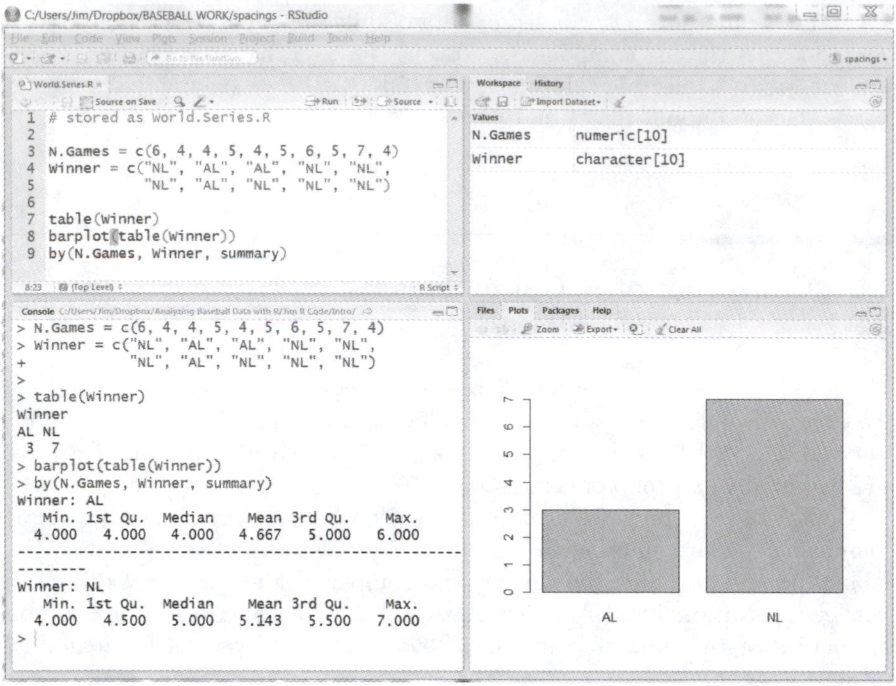

FIGURE 2.6
Snapshot of the RStudio interface after executing commands from an R script.

2.5.2   R functions

We have illustrated the use of a number of R built-in packages. One attractive feature of R is the capability to create one's own function to implement specific computations and graphs of interest.

As a simple example, suppose you are interested in writing a function to compute a player's home run rates for a collection of seasons. One inputs a vector `age` of player ages, a vector `hr` of home run counts, and a vector `ab` of at-bats. You want the function to compute the player's home run rates (as a percentage, rounded to the nearest tenth), and output the ages and rates in a form amiable for graphing.

The following function `hr.rates` will perform the desired calculations. All functions start with the syntax `Name.of.function <- function(arguments)`, where arguments is a list of input variables. All of the work in the function goes inside the curly brackets that follow. The result of the last line of the function is returned as the output. In our example, the name of the function is `hr.rates` and there are three vector inputs `age`, `hr`, and `ab`. The `round` function is used to compute the home run rates.[2] The output of this function

---

[2]The function `round(x, n)` rounds `x` to `n` decimal places.

is a list with two components: x is the vector of ages, and y is the vector of home run rates.

```
hr.rates <- function(age, hr, ab){
  rates <- round(100 * hr / ab, 1)
  list(x=age, y=rates)
}
```

To use this function, first it needs to be read into R. This can be done by entering it directly into the Console window, or by saving the function in a file, say hr.rates.R, and reading it into R by the source function.

```
source("hr.rates.R")
```

We illustrate using this function on some home run data for Mickey Mantle for the seasons 1951 to 1961. We enter Mantle's home run counts in the vector HR, the corresponding at-bats in AB, and the ages in Age. We apply the function hr.rates with inputs Age, HR, AB, and the output is a list with Mantle's ages and corresponding home run rates.

```
HR <- c(13, 23, 21, 27, 37, 52, 34, 42, 31, 40, 54)
AB <- c(341, 549, 461, 543, 517, 533, 474, 519, 541, 527, 514)
Age <- 19 : 29
hr.rates(Age, HR, AB)
$x
 [1] 19 20 21 22 23 24 25 26 27 28 29

$y
 [1]  3.8  4.2  4.6  5.0  7.2  9.8  7.2  8.1  5.7  7.6 10.5
```

One can easily construct a scatterplot (not shown here) of Mantle's rates against age by the plot function on the output of the function.

```
plot(hr.rates(Age, HR, AB))
```

Note that Mantle's home run rates rose steadily in the first six seasons of his career.

2.6   Reading and Writing Data in R

2.6.1   Importing data from a file

Generally it is tedious to input data manually into R. For the large data files that we will be working with in this book, it will be necessary to import these files directly into R. We illustrate this importing process using the complete pitching profile of Spahn.

We created the file "spahn.csv" containing Spahn's pitching statistics and placed the file in the current working directory. One can check the location of the current working directory in R by means of typing in the Console window:

```
getwd()
```

On RStudio, one can change the working directory by selecting the "Change Working Directory" option on the Tools menu or by use of the `setwd` function. One can easily import this dataset in RStudio by pressing the "Import Dataset" button in the top right window. You select the "From Text File" option and find the dataset of interest. After you select the file, Figure 2.7 shows a snapshot of the Import Dataset window. One sees the input file and also the format of the data that will be saved into R. It is important to check the button that the file contains a heading which means the first line of the input file contains the variable names.

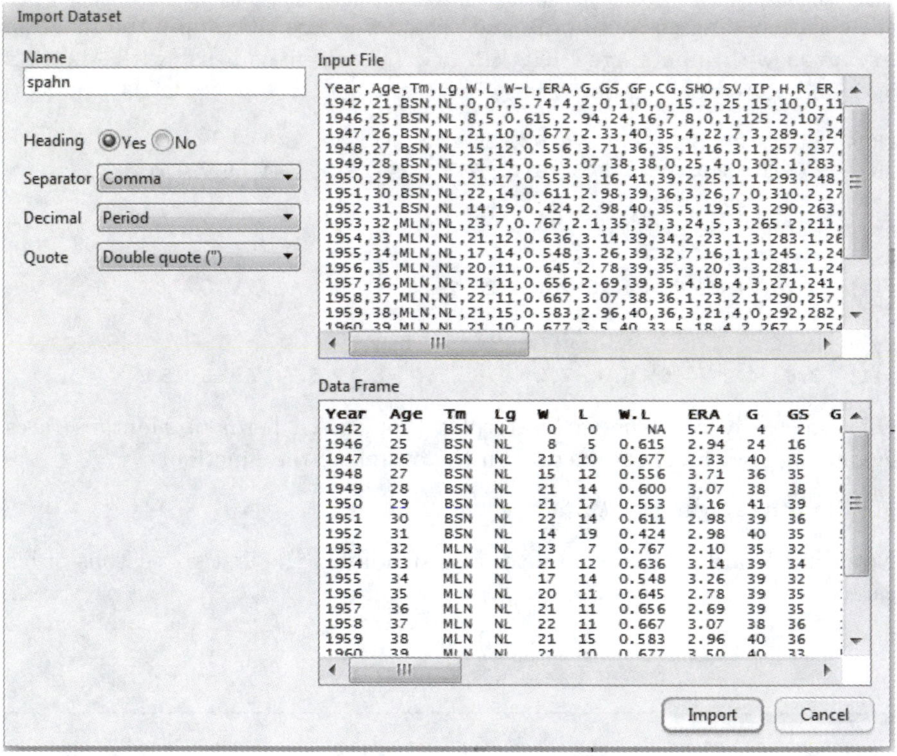

FIGURE 2.7
Snapshot of the Import Dataset window in the RStudio interface.

If the input file is in the current working directory, an alternative method of importing data from a file uses the `read.csv` function. This function assumes the file is stored in a "comma separated value" format, where different values on a single row are separed by commas. For our example, the following R expression reads the comma separated value file "spahn.csv" and saves the data into a data frame with name **spahn**.

```
spahn <- read.csv("spahn.csv")
```

2.6.2 Saving datasets

We have seen that it is straightforward to read comma-delimited data files (csv format) into R by use of the **read.csv** function. Similarly, we can use the **write.csv** function to save datasets in R in the csv format.

As an example, we return to the Mickey Mantle example where we have vectors of home run counts, at-bats, and ages and we use the user-defined function **hr.rates** to compute home run rates. We create a matrix **Mantle** attaching vectors **Age**, **HR**, **AB**, and the y component of the list **HR.Rates** using the **cbind** function.[3]

```
HR <- c(13, 23, 21, 27, 37, 52, 34, 42, 31, 40, 54)
AB <- c(341, 549, 461, 543, 517, 533, 474, 519, 541, 527, 514)
Age <- 19 : 29
HR.Rates <- hr.rates(Age, HR, AB)
Mantle <- cbind(Age, HR, AB, Rates=HR.Rates$y)
```

We use the **write.csv** function to save the data to the current working directory. This function has two arguments: the R object **Mantle** that we wish to save, and the output file name "mantle.csv". By using the argument **row.names = FALSE** option, row names will be omitted in the file that is saved.

```
write.csv(Mantle, "mantle.csv", row.names=FALSE)
```

It is good to confirm that a new file "mantle.csv" exists in the current working directory.

2.7 Data Frames

2.7.1 Introduction

The variable **spahn** described in the previous section is an example of a *data frame*, a fundamental object type in R, very similar to the table of pitching statistics that we saw on the back of Warren Spahn's baseball card in Figure 2.3. A data frame is a rectangular table of data, where rows of the table

---

[3]If one has three vectors a, b, c of equal lengths, the function cbind(a, b, c) combines the vectors into a matrix where the vectors are columns of the matrix.

correspond to different individuals or seasons, and columns of the table correspond to different variables collected on the individuals. Data variables can be numerical (like a batting average or a winning percentage), integer (like the count of home runs or number of wins), a factor (a categorical variable such as the player's team), or other types.

We can display portions of a data frame using the square bracket notation. For example, if we wish to display the first five rows and the first four variables (columns) of a data frame x, we type x[1 : 5, 1 : 4]. Below we display the first three rows and columns 1 though 10 of the spahn data frame.

```
spahn[1 : 3, 1 : 10]
  Year Age  Tm Lg  W  L   W.L  ERA  G GS
1 1942  21 BSN NL  0  0    NA 5.74  4  2
2 1946  25 BSN NL  8  5 0.615 2.94 24 16
3 1947  26 BSN NL 21 10 0.677 2.33 40 35
```

The header labels Year, Age, Tm, W, L, W.L, ERA, G, GS are some variable names of the data frame; the numbers 1, 2, 3 displayed on the left give the row numbers. We can display all variables for the first row by leaving the second argument blank.

```
spahn[1, ]
  Year Age  Tm Lg W L W.L  ERA G GS GF CG SHO SV   IP  H  R ER HR BB
1 1942  21 BSN NL 0 0  NA 5.74 4  2  0  1   0  0 15.2 25 15 10  0 11
  IBB SO HBP BK WP BF ERA.  WHIP  H.9 HR.9 BB.9 SO.9 SO.BB Awards
1  NA  7   0  0  0 79  59 2.298 14.4    0  6.3    4  0.64
```

A subset of the variables can be displayed by use of bracket notation. For example, if we wish to display the variables Age, W, L, ERA for the first 10 seasons, we would write the code

```
spahn[1 : 10, c("Age", "W", "L", "ERA")]
   Age  W  L  ERA
1   21  0  0 5.74
2   25  8  5 2.94
3   26 21 10 2.33
4   27 15 12 3.71
5   28 21 14 3.07
6   29 21 17 3.16
7   30 22 14 2.98
8   31 14 19 2.98
9   32 23  7 2.10
10  33 21 12 3.14
```

Individual variables of a data frame can be accessed by means of the $ notation. For example spahn$ERA would contain the earned run averages for Spahn's seasons. We can obtain some descriptive statistics for this measure by means of the summary function.

```
summary(spahn$ERA)
   Min. 1st Qu.  Median    Mean 3rd Qu.    Max.
  2.100   2.940   3.040   3.256   3.260   5.740
```

From this display, we see that 50% of Spahn's season eras fell between the lower quartile 2.940 and the upper quartile 3.260. Using logical operators, the age when Spahn had his lowest ERA can be found by use of the following expression.

```
spahn$Age[spahn$ERA == min(spahn$ERA)]
[1] 32
```

Using the ERA measure, Spahn had his best pitching season at the age of 32.

2.7.2 Manipulations with data frames

The pitching variables in the `spahn` data frame are the traditional or standard pitching statistics. One can add new "sabermetric" variables to the data frame by means of vector operations. Suppose that one wishes to measure pitching by the FIP (fielding independent pitching) statistic[4] defined by

$$FIP = \frac{13HR + 3BB - 2K}{IP}.$$

We add a new variable to a current data frame using the $ convention. In the following R code, the `with` function indicates that the variables HR, BB, SO, and IP are understood in the environment of the `spahn` data frame.

```
spahn$FIP <- with(spahn, (13 * HR + 3 * BB - 2 * SO) / IP)
```

Suppose we are interested in finding the seasons where Spahn performed the best using the FIP measure. The `order` function is used to give the positions of the measure where "1" corresponds to the smallest value; these positions are stored in the vector `pos`. By using the bracket notation, we sort the data frame using these positions. The `head` function is used to display only the first few rows of the data frame.

```
pos <- order(spahn$FIP)
head(spahn[pos, c("Year", "Age", "W", "L", "ERA", "FIP")])
   Year Age  W  L  ERA       FIP
8  1952  31 14 19 2.98 0.3448276
9  1953  32 23  7 2.10 0.3619910
2  1946  25  8  5 2.94 0.4153355
15 1959  38 21 15 2.96 0.6746575
3  1947  26 21 10 2.33 0.6950207
12 1956  35 20 11 2.78 0.8004269
```

---

[4]FIP is a measure of pitching performance dependent only on plays that do not involve fielders.

It is interesting that Spahn's best FIP seasons occurred during the middle of his career. Also, note that Spahn had a smaller (better) FIP in 1952 compared to 1953, although his ERA was significantly larger in 1952.

Since Spahn pitched primarily for two cities, Boston and Milwaukee, suppose we are interested in comparing his pitching for the two cities. We first create a new data frame spahn1 containing only the statistics for the two teams. This is done using the subset function with two arguments – the first argument is the data frame to subset, and the second argument is the logical condition defining the new data frame. (We introduce the logical OR operator |.)

```
spahn1 <- subset(spahn, Tm == "BSN" | Tm == "MLN")
```

The current factor variable Tm has three possible values, "BSN," "MLN," and "TOT" (for the total statistics for the 1965 season when Spahn played for two teams). We redefine Tm using the factor function so that there are only two possible values.

```
spahn1$Tm <- factor(spahn1$Tm, levels=c("BSN", "MLN"))
```

To compare various pitching statistics for the two teams, we use the by function. The three arguments to by are the data frame to be summarized (here the subset of the main data frame consisting of the variables W.L, ERA, WHIP, and FIP), the grouping variable Tm, and the function (here summary) that will be applied to each group. The output gives the summary statistics for the Boston seasons and the Milwaukee seasons.

```
by(spahn1[, c("W.L", "ERA", "WHIP", "FIP")], spahn1$Tm, summary)
spahn1$Tm: BSN
      W.L                ERA               WHIP              FIP
 Min.   :0.4240    Min.   :2.330    Min.   :1.136    Min.   :0.3448
 1st Qu.:0.5545    1st Qu.:2.970    1st Qu.:1.154    1st Qu.:0.6251
 Median :0.6000    Median :3.025    Median :1.222    Median :0.8219
 Mean   :0.5766    Mean   :3.364    Mean   :1.331    Mean   :0.7922
 3rd Qu.:0.6130    3rd Qu.:3.297    3rd Qu.:1.230    3rd Qu.:0.9836
 Max.   :0.6770    Max.   :5.740    Max.   :2.298    Max.   :1.2500
 NA's   :1
----------------------------------------------------------------
spahn1$Tm: MLN
      W.L                ERA               WHIP              FIP
 Min.   :0.3160    Min.   :2.100    Min.   :1.058    Min.   :0.3620
 1st Qu.:0.5780    1st Qu.:2.757    1st Qu.:1.123    1st Qu.:0.8345
 Median :0.6405    Median :3.030    Median :1.163    Median :0.9944
 Mean   :0.6202    Mean   :3.121    Mean   :1.187    Mean   :0.9839
 3rd Qu.:0.6695    3rd Qu.:3.170    3rd Qu.:1.226    3rd Qu.:1.0764
 Max.   :0.7670    Max.   :5.290    Max.   :1.474    Max.   :1.7263
```

It is interesting that Spahn's ERAs were higher in Boston (the middle 50% between 2.970 and 3.297 in Boston, compared to the middle 50% between

2.757 and 3.170 in Milwaukee), but Spahn's FIPs were lower in Boston. This indicates that Spahn may have had a weaker defense or unlucky with hits in balls in play in Boston.

2.7.3 Merging and selecting from data frames

In baseball research, it is common to have several data frames containing batting and pitching data for teams. Here we describe several ways of merging data frames and extract a portion of a data frame that satisfies a given condition.

Suppose we read into R data frames `NLbatting` and `ALbatting` containing batting statistics for all National League and American League teams in the 2011 season. Suppose we wish to merge or combine these data frames into a new data frame `batting`. To join two data frames vertically, we can use the `rbind` (for row combine) function.

```
NLbatting <- read.csv("NLbatting.csv")
ALbatting <- read.csv("ALbatting.csv")
batting <- rbind(NLbatting, ALbatting)
```

This command assumes that the two data frames `NLbatting` and `ALbatting` have the same variables; otherwise an error message will be displayed.

Suppose instead that we have read in the batting data `NLbatting` and the pitching data `NLpitching` for the NL teams in the 2011 season and we wish to merge these data frames horizontally. In this case, a row of the merged data frame would contain the batting and pitching statistics for a particular team. In this case, we use the function `merge` where we specify the two data frames and the `by` argument indicates the common variable (`Tm`) to merge by.

```
NLpitching <- read.csv("NLpitching.csv")
NL <- merge(NLbatting, NLpitching, by="Tm")
```

The new data frame `NL` contains 16 (the number of NL teams) rows and all of the variables from both the `NLbatting` and `ALbatting` data frames.

A third useful operation is choosing a subset of a data frame that satisfies a particular condition. Suppose one has the data frame `NLbatting` and one wishes to focus on the batting statistics for only the teams who hit over 150 home runs this season. We use the `subset` function – the arguments are the original data frame and the logical condition that describes how teams are selected.

```
NL.150 <- subset(NLbatting, HR > 150)
```

The new data frame `NL.150` contains the batting statistics for the eight teams who hit over 150 home runs.

2.8   Packages

This book will focus on functions available on the base R system that is
installed. But one attractive feature of R is the availability of collections of
functions and datasets in R packages. Currently, there are over 4000 packages
contributed by R users available on the R website `cran.r-project.org/` and
these packages expand the capabilities of the R system. In our book, we focus
on a few contributed packages that we find useful in our baseball work.

To illustrate installing and loading an R package, we recently found a package **Lahman**, that contains the data files from the Lahman database described
in Section 1.2. Assuming one is connected to the Internet, one can install the
current version of this package into R by means of the command

```
install.packages("Lahman")
```

Alternately, one can install packages by use of the Install Packages button on
the Package tab in RStudio.

After a package has been installed, then one needs to load the package into
R to have access to the functions and datasets. For example, to load the new
package **Lahman**, one types

```
library(Lahman)
```

To confirm that the package has been loaded correctly, we use the `help` function to learn about the dataset `Batting` in the **Lahman** package. (A general
discussion of the `help` function is given in Section 2.10.)

```
?Batting
```

When one launches R, one needs to load the packages that are not automatically loaded in the system.

2.9   Splitting, Applying, and Combining Data

In many situations, one is interested in splitting a data frame into parts,
applying some operation on each part, and then combining the results in a
new data frame. This type of "split, apply, combine" operation is facilitated
using the function `sapply` and special functions in the `plyr` package. Here we
illustrate this process on the Lahman batting database. In this work, we review
some other handy data frame manipulation functions previously discussed.

2.9.1   Using `sapply`

Suppose we are interested in looking at the great home run hitters in baseball history. Specifically, we want to answer the question "Who hit the most home runs in the 1960s?"

We begin by reading in Lahman's batting data file "Batting.csv" containing the season batting statistics for all players in baseball history – the data frame is stored in the variable `Batting`.

```
Batting <- read.csv("Batting.csv")
```

Since we are focusing on the 1960s, the `subset` function is used to select batting data only for the seasons between 1960 and 1969, creating the new data frame `Batting.60`.

```
Batting.60 <- subset(Batting, yearID >= 1960 & yearID <= 1969)
```

The `sapply` function is useful for repeating a particular operation over a set of values in a vector. In this example, we would like to compute the total number of home runs for each player in the data frame `Batting.60`. First, a function `compute.hr` is written that computes the total home runs for a player with `playerID` equal to `pid`.

```
compute.hr <- function(pid){
  d <- subset(Batting.60, playerID == pid)
  sum(d$HR)
}
```

By use of the `unique` function, a vector of the ids for all of the players in the 1960s is created. The home runs for all players is accomplished by the `sapply` function – the arguments are the vector `players` and the function `compute.hr` that will be applied to each element in the vector. The output is a vector S containing the total home run count for all players in the vector.

```
players <- unique(Batting.60$playerID)
S <- sapply(players, compute.hr)
```

A new data frame R is created using the `data.frame` function. The syntax indicates there are two variables in the new data frame – **Player** correspond to the player ids contained in the vector `players` and HR corresponds to the home run counts contained in the vector S. Using the `order` function, we sort this data frame so that the best home run hitters are on the top, and display the first lines of this data frame.

```
R <- data.frame(Player=players, HR=S)
R <- R[order(R$HR, decreasing=TRUE), ]
head(R)
        players    S
857   killeha01  393
1     aaronha01  375
```

```
1045  mayswi01 350
1373 robinfr02 316
1058 mccovwi01 300
752  howarfr01 288
```

The best home run hitters in the 1960s were Harmon Killebrew, Hank Aaron, Willie Mays, and Frank Robinson.

2.9.2   Using `ddply` in the `plyr` package

The `plyr` package provides a more extensive collection of tools for this "split, apply, combine" operation. Using the same `Batting` data frame of season batting statistics, suppose we are interested in collecting the career at-bats, career home runs, and career strikeouts for all players in baseball history with at least 5000 career at-bats. Both home runs and strikeouts are of interest since we suspect there may be some association between a player's strikeout rate (defined by SO/AB) and his home run rate HR/AB.

A new data frame consisting of hitting data of batters with at least 5000 career at-bats will be created. This operation is done in three steps. First, a new data frame is created consisting of the career AB for all batters. Second, this new data frame is merged with the original data frame `Batting`, creating a new variable `Career.AB`. Last, by use of the `subset` function, the batting seasons are selected from the data frame for the players with 5000 AB.

The function `ddply` in the `plyr` package is useful for the first operation. We want to compute the sum of AB over the seasons of a player's career. There are three arguments to the `ddply` function. `Batting` is the data frame we wish to split, `.(player.ID)` indicates we wish to split the data frame by the player id variable, and `summarize, Career.AB=sum(AB, na.rm=TRUE)` indicates we wish to summarize each data frame "part" by computing the sum of the AB. (Some of the AB values will be missing and coded as "NA", and the `na.rm=TRUE` will remove these missing values before taking the sum.) The new data frame `dataframe.AB` contains the career AB for all players.

```
library(plyr)
dataframe.AB <- ddply(Batting, .(playerID), summarize,
          Career.AB=sum(AB, na.rm=TRUE))
```

We want to add a new variable `Career.AB` to the original data frame. This is done by use of the `merge` function, merging data frames `Batting` and `dataframe.AB`, matching by the common variable `playerID`.

```
Batting <- merge(Batting, dataframe.AB, by="playerID")
```

Now that we have this new variable `Career.AB`, one can now use the `subset` function to choose only the season batting statistics for the players with 5000 AB.

```
Batting.5000 <- subset(Batting, Career.AB >= 5000)
```

For each player in the data frame `Batting.5000`, we want to compute the career AB, career HR, and career SO. This is another example of the "split, apply, combine" operation done conveniently by the `ddply` function. We first write a small function `ab.hr.so` that works on the season batting statistics for a single player. The input is the data frame `d` containing the statistics for one player and the output is a data frame with the career AB, HR, and SO.

```
ab.hr.so <- function(d){
  c.AB <- sum(d$AB, na.rm=TRUE)
  c.HR <- sum(d$HR, na.rm=TRUE)
  c.SO <- sum(d$SO, na.rm=TRUE)
  data.frame(AB=c.AB, HR=c.HR, SO=c.SO)
}
```

To illustrate the use of `ab.hr.so`, we extract Hank Aaron's batting statistics and apply this function on Aaron's data frame `aaron`.

```
aaron <- subset(Batting.5000, playerID == "aaronha01")
ab.hr.so(aaron)
     AB  HR   SO
1 12364 755 1383
```

This confirms that Aaron had 755 career home runs and 1383 career strikeouts.

To apply this function to each batter and collect the results, we again use the function `ddply`. The arguments are the data frame `Batting.5000` to split, the splitting variable `player.ID`, and the function `ab.hr.so` to apply on each part.

```
d.5000 <- ddply(Batting.5000, .(playerID), ab.hr.so)
```

The resulting data frame `d.5000` contains the career AB, HR, and SO for all batters with at least 5000 career AB. To confirm, the first six lines of the data frame are displayed by the `head` function.

```
head(d.5000)
   playerID    AB  HR   SO
1 aaronha01 12364 755 1383
2 abreubo01  8128 284 1763
3 adamssp01  5557   9  223
4 adcocjo01  6606 336 1059
5 alfoned01  5385 146  617
6 allendi01  6332 351 1556
```

Is there an association between a player's home run rate and his strike-out rate? Using the `plot` function, we construct a scatterplot of `HR/AB` and `SO/AB`. Using the `lines` function we add a smoothing curve (using the function `lowess`) to the scatterplot. (See Figure 2.8.)

```
with(d.5000, plot(HR/AB, SO/AB))
with(d.5000, lines(lowess(HR/AB, SO/AB)))
```

It is clear from the graph that batters with higher home run rates tend to have higher strikeout rates.

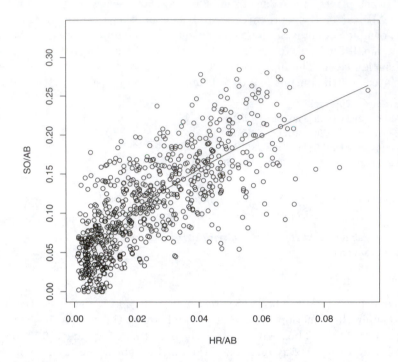

FIGURE 2.8
Scatterplot of the homerun rates and strikeout rates of all players with at least 5000 career at-bats. A smoothing curve is added to the plot to show that home run rates and strikeout rates have a positive association.

2.10   Getting Help

The Help menu in RStudio provides general documentation about the R system (see the R Help option). From the Help menu, we also find general information about the RStudio system such as keyboard shortcuts. In addition, R contains an online help system providing documentation on R functions and datasets. For example, suppose you wish to learn about the `dotchart` function which constructs a dot chart, a new type of statistical graphical display discussed in Chapter 3. By typing in the Console window a question mark followed by the function name,

```
?dotchart
```

you see a long description of this function including all of the possible function arguments. To find out about related functions, one can preface "dotchart" by two question marks to find all objects that contain this character string:

```
??dotchart
```

On my system, I learn about the function `dotchart2` in the `Hmisc` package that constructs an enhanced dot chart display.

RStudio provides an additional online help system which is especially helpful when one does not know the exact spelling of an R function. For example, suppose I want to construct a dot chart, but all I know is that the function contains the string "dot". In the Console window, I type "dot" followed with a Tab. RStudio will complete the code, forming "dotplot" and show an abbreviated description of the function. In the case where the character string does not uniquely define the function, RStudio will display all of the functions with that string.

2.11 Further Reading

R is an increasingly popular system for performing data analysis and graphics and a large number of books are available which introduce the system. The manual "An Introduction to R" (Venables et al., 2011) available on the R and RStudio systems provides a broad overview of the R language and the manual "R Data Import/Export" provides an extended description of R capabilities to import and export datasets. Kabacoff (2011) and the accompanying website `www.statmethods.net` provide helpful advice on specific R functions on data input, data management, and graphics. Albert and Rizzo (2012) provide an example-based introduction to R, where different chapters are devoted to specific statistics topics such as exploratory fitting, modeling, graphics, and simulation.

2.12 Exercises

1. (**Top Base Stealers in the Hall of Fame**)
   The following table gives the number of stolen bases (SB), the number of times caught stealing (CS), and the number of games played (G) for nine players currently inducted in the Hall of Fame.

Player	SB	CS	G
Rickey Henderson	1406	335	3081
Lou Brock	938	307	2616
Ty Cobb	897	212	3034
Eddie Collins	741	195	2826
Max Carey	738	109	2476
Joe Morgan	689	162	2649
Luis Aparicio	506	136	2599
Paul Molitor	504	131	2683
Roberto Alomar	474	114	2379

(a) In R, place the stolen base, caught stealing, and game counts in the vectors SB, CS, and G.

(b) For all players, compute the number of stolen base attempts SB + CS and store in the vector SB.Attempt.

(c) For all players, compute the success rate Success.Rate = SB / SB.Attempt.

(d) Compute the number of stolen bases per game SB.Game = SB / Game.

(e) Construct a scatterplot of the stolen bases per game against the success rates. Are there particular players with unusually high or low stolen base success rates? Which player had the greatest number of stolen bases per game?

2. (**Character, Factor, and Logical Variables in R**)

Suppose one records the outcomes of a batter in ten plate appearances:

Single, Out, Out, Single, Out, Double, Out, Walk, Out, Single

(a) Use the c function to collect these outcomes in a character vector outcomes.

(b) Use the table function to construct a frequency table of outcomes.

(c) In tabulating these results, suppose one prefers the results to be ordered from least-successful to most-successful. Use the following code to convert the character vector outcomes to a factor variable f.outcomes.

```
f.outcomes <- factor(outcomes,
    levels=c("Out", "Walk", "Single", "Double"))
```

Use the table function to tabulate the values in f.outcomes. How does the output differ from what you saw in part (b)?

(d) Suppose you want to focus only on the walks in the plate appearances. Describe what is done in each of the following statements.

```
outcomes == "Walk"
sum(outcomes == "Walk")
```

3. (**Pitchers in the 350-Wins Club**)

The following table lists nine pitchers who have won at least 350 career wins.

Player	W	L	SO	BB
Pete Alexander	373	208	2198	951
Roger Clemens	354	184	4672	1580
Pud Galvin	364	310	1806	745
Walter Johnson	417	279	3509	1363
Greg Maddux	355	227	3371	999
Christy Mathewson	373	188	2502	844
Kid Nichols	361	208	1868	1268
Warren Spahn	363	245	2583	1434
Cy Young	511	316	2803	1217

(a) In R, place the wins and losses in the vectors W and L, respectively. Also, create a character vector Name containing the last names of these pitchers.

(b) Compute the winning percentage for all pitchers defined by $100 \times W/(W+L)$ and put these winning percentages in the vector Win.PCT.

(c) By use of the command

```
Wins.350 <- data.frame(Name, W, L, Win.PCT)
```

create a data frame Wins.350 containing the names, wins, losses, and winning percentages.

(d) By use of the order function, sort the data frame Wins.350 by winning percentage. Among these pitchers, who had the largest and smallest winning percentages?

4. (**Pitchers in the 350-Wins Club, Continued**)

(a) In R, place the strikeout and walk totals from the 350 win pitchers in the vectors SO and BB, respectively. Also, create a character vector Name containing the last names of these pitchers.

(b) Compute the strikeout-walk ratio by SO/BB and put these ratios in the vector SO.BB.Ratio.

(c) By use of the command

```
SO.BB <- data.frame(Name, SO, BB, SO.BB.Ratio)
```

create a data frame SO.BB containing the names, strikeouts, walks, and strikeout-walk ratios.

(d) By use of the subset function, find the pitchers who had a strikeout-walk ratio exceeding 2.8.

(e) By use of the `order` function, sort the data frame by the number of walks. Did the pitcher with the largest number of walks have a high or low strikeout-walk ratio?

5. (**Pitcher Strikeout/Walk Ratios**)

(a) Read the Lahman "pitching.csv" data file into R into a data frame `Pitching`.

(b) The following function computes the cumulative strikeouts, cumulative walks, mid career year, and the total innings pitched (measured in terms of outs) for a pitcher whose season statistics are stored in the data frame d.

```
stats <- function(d){
  c.SO <- sum(d$SO, na.rm=TRUE)
  c.BB <- sum(d$BB, na.rm=TRUE)
  c.IPouts <- sum(d$IPouts, na.rm=TRUE)
  c.midYear <- median(d$yearID, na.rm=TRUE)
  data.frame(SO=c.SO, BB=c.BB, IPouts=c.IPouts,
             midYear=c.midYear)
}
```

Using the function `ddply` (`plyr` package) together with the function `stats`, find the career statistics for all pitchers in the pitching dataset. Call this new data frame `career.pitching`.

(c) Use the `merge` function to merge the `Pitching` and `career.pitching` data frames.

(d) Use the `subset` function to construct a new data frame `career.10000` consisting of data for only those pitchers with at least 10,000 career IPouts.

(e) For the pitchers with at least 10,000 career IPouts, construct a scatterplot of mid career year and ratio of strikeouts to walks. Comment on the general pattern in this scatterplot.

3

Traditional Graphics

CONTENTS

3.1	Introduction	59
3.2	Factor Variable	60
	3.2.1 A bar graph	60
	3.2.2 Add axes labels and a title	61
	3.2.3 Other graphs of a factor	62
3.3	Saving Graphs	62
3.4	Dot plots	63
3.5	Numeric Variable: Stripchart and Histogram	65
3.6	Two Numeric Variables	67
	3.6.1 Scatterplot	67
	3.6.2 Building a graph, step-by-step	69
3.7	A Numeric Variable and a Factor Variable	73
	3.7.1 Parallel stripcharts	74
	3.7.2 Parallel boxplots	74
3.8	Comparing Ruth, Aaron, Bonds, and A-Rod	76
	3.8.1 Getting the data	76
	3.8.2 Creating the player data frames	78
	3.8.3 Constructing the graph	78
3.9	The 1998 Home Run Race	79
	3.9.1 Getting the data	79
	3.9.2 Extracting the variables	81
	3.9.3 Constructing the graph	82
3.10	Further Reading	82
3.11	Exercises	83

3.1 Introduction

To illustrate basic methods for creating graphs in R in the `graphics` package, consider all the career batting statistics for the current members of the Hall of Fame. If we remove the pitchers' batting statistics from the dataset, then we have statistics for 147 non-pitchers. The data file "hofbatting.csv" contains the career batting statistics for this group. We read this data file into R by the `read.csv` function; the statistics are stored in a data frame named `hof`.

```
hof <- read.csv("hofbatting.csv")
```

The type of graph we use depends on the measurement scale of the variable. There are two fundamental data types – measurement and categorical – which are represented in R as numeric and factor variables. We initially describe graphs for a single factor variable and a single numeric variable, and then describe graphical displays helpful for understanding relationships between the variables. Using the traditional graphics package of R, it is easy to modify the attributes of a graph by adding labels and changing the style of plotting symbols and lines. After describing the graphical methods, we describe the process of creating graphs for two home run stories. The first graph compares the home run career progress of four great sluggers in baseball history and the second graph illustrates the famous home run race of Mark McGwire and Sammy Sosa during the 1998 season.

3.2   Factor Variable

3.2.1   A bar graph

The Hall-of-Famers played during different eras of baseball; one common classification of eras is "19th Century" (up to the 1900 season), "Dead Ball" (1901 through 1919), "Lively Ball" (1920 though 1941), "Integration" (1942 through 1960), "Expansion" (1961 through 1976), "Free Agency" (1977 through 1993), and "Long Ball" (after 1993). We want to create a new factor variable `Era` giving the era for each player. First a player's mid career (variable `MidCareer`) is defined as the average of his first and last seasons in baseball. The `cut` function creates the new factor variable `Era` – the arguments of the function are the numeric variable to be converted, the vector of cut points, and the vector of labels for the categories of the factor variable.

```
hof <- read.csv("hofbatting.csv")
hof$MidCareer <- with(hof, (From + To) / 2)
hof$Era <- cut(hof$MidCareer,
        breaks = c(1800, 1900, 1919, 1941, 1960, 1976, 1993, 2050),
        labels = c("19th Century", "Dead Ball", "Lively Ball",
                   "Integration", "Expansion", "Free Agency",
                   "Long Ball"))
```

A bar graph of a factor variable is constructed using the `barplot` function. We first construct a frequency table of the variable `Era` using the `table` function and store the output into the variable `T.Era`. The output of `table` is the input for `barplot`. Figure 3.1 (a) shows the resulting graph. We see that a large number of these Hall of Fame players played during the Lively Ball era.

```
T.Era <- table(hof$Era)
T.Era
```

19th Century	Dead Ball	Lively Ball	Integration	Expansion
17	19	46	24	21
Free Agency	Long Ball			
18	2			

```
barplot(T.Era)
```

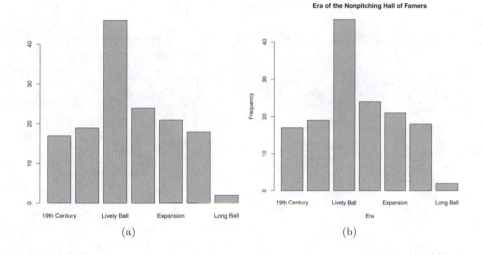

(a) (b)

FIGURE 3.1
Bar graphs of era of the Hall of Fame non-pitchers. The right graph adds axes labels and a title to the basic plot.

3.2.2   Add axes labels and a title

As good practice, graphs should have descriptive axes labels and a title for describing the main message of the display. In the traditional graphics displays in R, the arguments `xlab` and `ylab` add horizontal and vertical axis labels and the `main` argument adds a title. In the following `barplot` function, we add the labels "Era" and "Frequency" and add the title "Era of the Nonpitching Hall of Famers." The enhanced plot is shown in Figure 3.1(b).

```
barplot(table(hof$Era), xlab="Era", ylab="Frequency",
        main="Era of the Nonpitching Hall of Famers")
```

3.2.3   Other graphs of a factor

There are alternative graphical displays for a table of frequencies of a factor variable. For the table of era frequencies, the function

```
plot(table(hof$Era))
```

draws a vertical line graph of the frequencies and the function

```
pie(table(hof$Era))
```

constructs a pie chart. Figure 3.2 shows these alternative displays A line graph

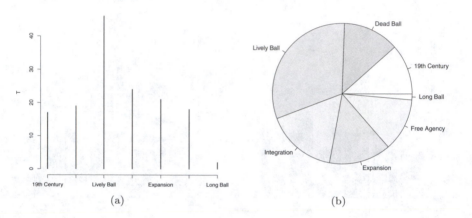

(a)   (b)

FIGURE 3.2
Line graph and pie chart of the frequencies of era of the Hall of Fame non-pitchers.

is helpful when there are a large number of categories of the factor. Although pie charts are popular in displaying frequencies in the media, we prefer a bar graph since it can be more difficult for a reader to visually compare the relative sizes of slices of a pie chart than lengths of bars in a bar chart.

3.3   Saving Graphs

After a graph is produced in R, it is straightforward to export the graph to one of the the usual graphics formats so that it can be used in a document, blog, or website. We outline the steps for saving graphs in the RStudio interface.

   If a graph appears in the Plots window of RStudio, then the Export menu allows one to "Save Plot as Image," "Save Plot as PDF," or "Copy Plot to the Clipboard." If one chooses the "Save Plot as Image" option, then by choosing an option from a drop-down menu, one can save the graph in PNG, JPEG, TIFF, BMP, metafile, clipboard, SVG, or EPS formats. The PNG format is convenient for uploading to a web page and the EPS and PDF formats are well-suited for use in a LaTex document. The metafile and clipboard options are useful for insertion of the graph into a Microsoft Word document.

Alternately, plots can be saved by use of R functions typed in the Console window. For example, suppose we wish to save the bar graph shown in Figure 3.1(b) in a graphics file of PNG format. We use the special **png** function where the argument is the name of the saved graphics file. We follow this with the R commands to produce the graph, and conclude with the **dev.off** function. (Note that no graph will be displayed in this operation.)

```
png("bargraph.png")
barplot(table(hof$Era), xlab="Era", ylab="Frequency",
  main="Era of the Nonpitching Hall of Famers")
dev.off()
RStudioGD
        2
```

If we look at the current directory, we will see a new file "bargraph.png" containing the image in PNG format. If one types the help command

```
?png
```

one will see command instructions for saving graphs in other graphics formats. This method of saving graphs is especially useful if one wishes to save a number of graphs in a single file. For example, if one types

```
pdf("graphs.pdf")
barplot(table(hof$Era))
plot(table(hof$Era))
dev.off()
RStudioGD
        2
```

then the bar graph and the line graph will be saved together in the PDF file "graphs.pdf."

3.4 Dot plots

A modern method of displaying labeled data is a dot plot, introduced by Cleveland (1994). This is a useful alternative to a bar graph, and is nice for displaying statistics for a group of baseball players.

We revisit the example where we are exploring the era for the Hall of Fame non-pitchers. The table of era frequencies is stored in the variable `T.Era`. We first convert `T.Era` to a numeric variable by use of the `as.numeric` function. To construct a dot plot, we use the `dotchart` function – the first argument is the vector of numeric values to display and the `label` argument is a vector of the corresponding labels.

```
T.Era <- table(hof$Era)
dotchart(as.numeric(T.Era), labels=names(T.Era), xlab="Frequency")
```

The dot plot, shown in Figure 3.3 simply represents each frequency by an open circle on a scale against the corresponding label. It is easy to compare the era frequencies from this display.

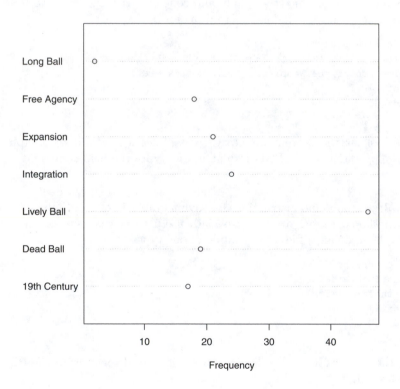

FIGURE 3.3
Dot plot of era of the Hall of Fame non-pitchers.

Dot plots can be used to display any collection of labeled numeric data. Suppose we are interested in exploring the career OPS values for the Hall of Fame players with at least 500 career home runs. We first use the `subset` function to create a new data frame `hof.500` consisting of the statistics of the players with at least 500 career home runs. By use of the `order` function, we order the rows of this matrix by the career OPS. We use the `dotchart` function to construct a dot plot of the OPS values where the labels are the names of the players (variable X in the data frame hof.500).

```
hof.500 <- subset(hof, HR >= 500)
hof.500 <- hof.500[order(hof.500$OPS), ]
dotchart(hof.500$OPS, labels=hof.500$X, xlab="OPS")
```

In this display (Figure 3.4), we see that Babe Ruth, Ted Williams, and Jimmie Foxx stand out as the top OPS players in this 500-home run group.

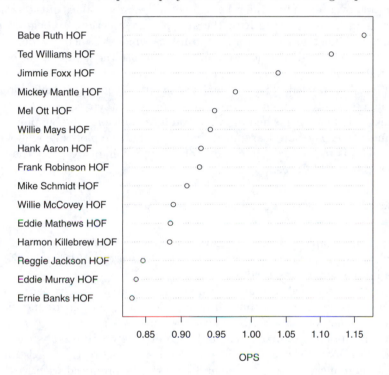

FIGURE 3.4
Dot plot of career OPS values for Hall of Famers with at least 500 career home runs.

3.5 Numeric Variable: Stripchart and Histogram

When one collects a numeric variable such as a batting average, an on-base percentage, or an OPS from a group of players, one typically wants to learn about its distribution. For example, if we examine OPS values for the non-pitcher HOF inductees, we are interested in learning about the general shape of the OPS values. For example, is the distribution of OPS values symmetric, or is the distribution right or left skewed? Also we are interested in learning about the typical or representative Hall-of-Fame OPS value, and how the OPS values are spread out. Graphical displays provide a quick visual way of studying distributions of collections of baseball statistics.

For a single numeric variable, two useful displays for visualizing a distribution are the stripchart or one-dimensional scatterplot, and the histogram. A stripchart is basically a number line graph, where the values of the statistics are plotted over a number line ranging over all possible values of the variable. This graph is constructed in R by the `stripchart` function. To begin, we use the `windows` function to open a new graphics window 7 inches wide and 3.5 inches tall.[1] This is done since we don't want to use the default 7 in. by 7 in. format. The only required argument in the `stripchart` function is the vector of data to be graphed. The optional argument `method="jitter"` indicates that the points are randomly placed in a band over their values; this "jittering" method of plotting is helpful when you have multiple plotting points with the same value. The `pch=1` argument indicates that the plotting symbol 1 (an open circle) is to be used,[2] and the `xlab` argument indicates that the x axis is labeled by "Mid Career".

```
windows(width=7, height=3.5)
stripchart(hof$MidCareer, method="jitter", pch=1,
           xlab="Mid Career")
```

The resulting graph is shown in Figure 3.5. One interesting observation from this graph is the presence of a gap between the seasons 1910 and 1920 with no Hall of Famers represented.

A second graphical display for a numeric variable is a histogram where the values are grouped into bins of equal width and the bin frequencies are displayed as non-overlapping bars over the bins. A histogram is constructed in R using the function `hist`. The only required input to `hist` is the vector of mid careers `hof$MidCareer`. The `xlab` adds a label to the x axis and the `main=""` argument removes the default title that is produced with `hist`.

```
hist(hof$MidCareer, xlab="Mid Career", main="")
```

The histogram of mid career values in Figure 3.6, as expected, resembles the bar graph of the variable `Era` described in the previous section. One issue in constructing a histogram is the choice of bins and the function `hist` will typically make reasonable choices for the bins to produce a good display of the data distribution. One can select one's own bins in the function `hist` by use of the argument `breaks`. For example, if one wanted to choose the alternative bin endpoints 1880, 1900, 1920, 1940, 1960, 1980, 2000, then one could construct the histogram by the following code (the figure is not displayed):

---

[1]On a Macintosh computer, the `quartz` function is used to open up a new graphics window.

[2]See `www.statmethods.net/advgraphs/parameters.html` for a display of all the possible plotting symbols possible with the `pch` argument.

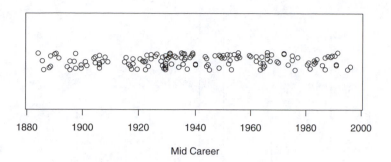

Mid Career

FIGURE 3.5
One-dimensional scatterplot (stripchart) of mid career values of HOF non-pitchers.

```
hist(hof$MidCareer, xlab="Mid Career", main="",
        breaks=seq(1880, 2000, by=20))
```

3.6 Two Numeric Variables

3.6.1 Scatterplot

When one collects two numeric variables for many players, one is interested in exploring their relationship. A scatterplot is a standard method for graphing two numeric variables and the workhorse function for constructing a scatterplot in R is `plot`.

A good measure of batting performance is the OPS statistic, the sum of the on-base percentage (OBP) and the slugging percentage (SLG). Is there any relationship between a player's OPS and the baseball era? Were there particular seasons where the HOF OPS values were unusually high or low?

We can answer these questions by constructing a scatterplot of the variables `MidCareer` and `OPS`. The `plot` function has two arguments, the variable to be plotted on the horizontal scale and the variable on the vertical scale. As it can be difficult to visually detect scatterplot patterns, it is helpful to add a smoothing curve to show the general association pattern. The popular loess smoothing method (Cleveland, 1979) is implemented by the `lowess` function with three arguments: the two variables and a constant `f` between 0 and 1 indicating the degree of smoothness in the curve. (Larger values of `f` result in smoother curves with fewer wrinkles.) By using the `lines` function, we add this smoothing curve to the current scatterplot. In viewing the

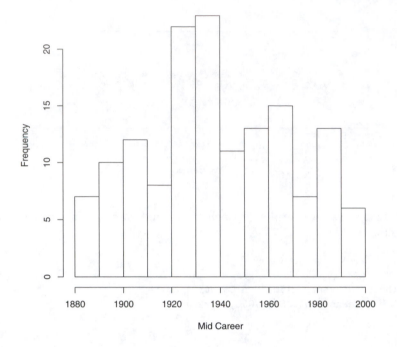

FIGURE 3.6
Histogram of mid career values of HOF nonpitchers.

scatterplot (see Figure 3.7), we notice four unusual career OPS values, three large values and one very small value, and we'd like to identify the players with these extreme values. Identification of specific points is accomplished by the `identify` function; the inputs are the two variables in the scatterplot, the vector giving labels (here, names) for the points, and the number of points to identify. When `identify` is executed, a hairline will appear when the mouse is placed on the graph, and the user clicks the mouse near the extreme points. Figure 3.7 shows the scatterplot with points identified.

```
with(hof, plot(MidCareer, OPS))
with(hof, lines(lowess(MidCareer, OPS, f=0.3)))
with(hof, identify(MidCareer, OPS, X, n=4))
```

What do we learn from Figure 3.7? The typical OPS of a Hall of Famer has stayed pretty constant through the years. But there was an increase in the OPS during the 1930s when Babe Ruth and Lou Gehrig were in their primes. There has been a steady decline in the average OPS (among these Hall of Famers) over the least 30 years. It is interesting to note that the variability of the OPS values among these players seems small in recent seasons.

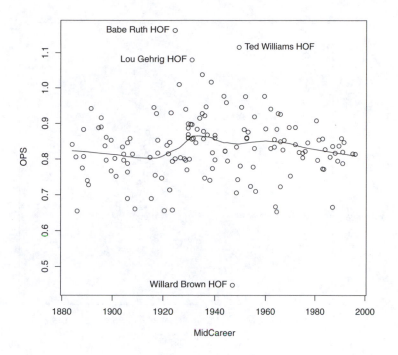

FIGURE 3.7
Scatterplot of mid career against season for HOF nonpitchers.

3.6.2 Building a graph, step-by-step

Generally, constructing a graph is an iterative process. One begins by choosing variables of interest and a particular graphical method (such as a scatterplot). By inspecting the resulting display, one will typically find ways for the graph to be improved. By using several of the optional arguments, one can make changes to the graph that result in a clearer and more informative display. We illustrate this graph construction process in the situation where one is exploring the relationship between two variables.

There are two dimensions of hitting, the ability to get on base, measured by the on-base percentage OBP, and the ability to advance runners already on base, measured by the slugging percentage SLG. One can better understand the hitting performances of players by constructing a scatterplot of these two measures. We use the `plot` function to construct a scatterplot of OBP and SLG. (See Figure 3.8.)

```
with(hof, plot(OBP, SLG))
```

Looking at this figure, we see several problems with this display. First, due to the one outlier in the bottom-left section of the graph, most of the

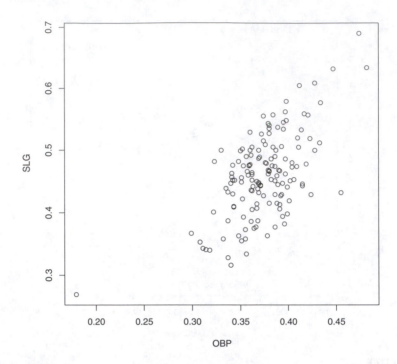

FIGURE 3.8
Scatterplot of OBP against SLG for HOF members.

points fall in a relatively small region of the plotting region. Second, it may
be preferable to use an alternative plotting symbol such as a filled circle that
is more distinctive than the default open circle symbol. Last, the graph would
be easier to read if more descriptive labels were used for the two axes. A new
figure is plotted to incorporate these new ideas. By use of the `xlim` and `ylim`
arguments, we change the limits of the horizontal and vertical axes. By the
new choices of the limits (0.28, 0.50) for the horizontal and (0.28, 0.75) for
the vertical, we remove the outlier and allow for more space in the upper-left
section of the graph for labels. By use of the `pch=19` argument, we change the
plotting symbol to a sold black circle. We use the `xlab` and `ylab` arguments to
replace OBP and SLG respectively with "On-Base Percentage" and "Slugging
Percentage." The updated display is shown in Figure 3.9.

```
with(hof, plot(OBP, SLG, xlim=c(0.25, 0.50),
              ylim=c(0.28, 0.75), pch=19,
              xlab="On-Base Percentage",
              ylab="Slugging Percentage"))
```

A good measure of batting performance is the OPS statistic defined by

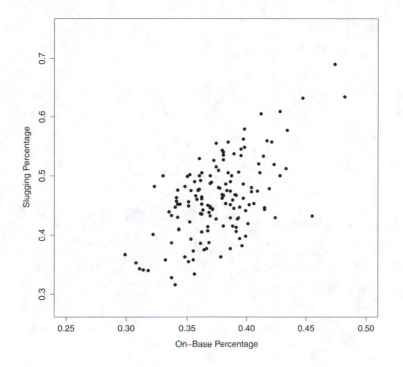

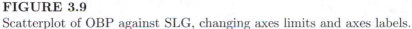

FIGURE 3.9
Scatterplot of OBP against SLG, changing axes limits and axes labels.

OPS = OBP + SLG. To evaluate hitters in our graph on the basis of OPS, it would be helpful to draw constant values of OPS on the graph. If we represent OBP and SLG by x and y, suppose we wish to draw a line where OPS = 0.7 or where $x+y = 0.7$. Equivalently, we want to draw the function $y = 0.7-x$ on the graph; this is accomplished in R by the `curve` function where the argument of the function is represented by the variable x. The `add=TRUE` arguments indicate that this function is to be drawn on the current graph. Similarly, we apply the `curve` function three more times to draw lines on the graph where OPS takes on the values 0.8, 0.9, and 1.0. The resulting display is shown in Figure 3.10.

```
curve(.7 - x, add=TRUE)
curve(.8 - x, add =TRUE)
curve(.9 - x, add=TRUE)
curve(1.0 - x, add=TRUE)
```

In our final iteration, we add labels to the lines showing the constant values of OPS, and we label the points corresponding to players having a lifetime OPS exceeding one. Each of the line labels is accomplished using the `text`

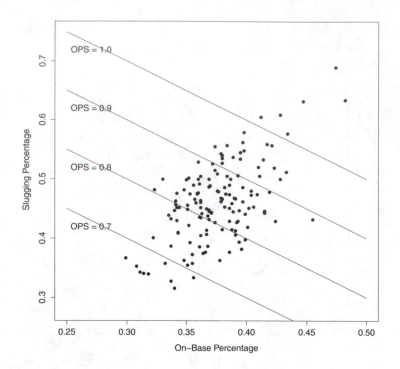

FIGURE 3.10
Scatterplot of OBP against SLG, adding lines of constant values of OPS =
OBP + SLG.

function – the three arguments are the x location and y location where the
text is to be drawn, and the string of text to be displayed.

```
text(.27, .42, "OPS = 0.7")
text(.27, .52, "OPS = 0.8")
text(.27, .62, "OPS = 0.9")
text(.27, .72, "OPS = 1.0")
```

To label the points for the best hitters by use of mouse clicks, the `identify`
function is used. The inputs are the x and y plotting variables, the vector of
point labels, and the number of points to label. The final graph is displayed
in Figure 3.11.

```
with(hof, identify(OBP, SLG, X, n=6))
```

This final graph is very informative about the batting performance of these
Hall of Famers. We see that a large group of these batters have career OPS
values between 0.8 and 0.9, and only six players (Hank Greenberg, Roger
Hornsby, Jimmie Foxx, Ted Williams, Lou Gehrig, and Babe Ruth) had career OPS values exceeding 1.0. Points to the right of the major point cloud

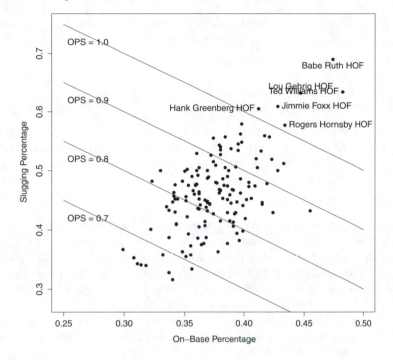

FIGURE 3.11
Scatterplot of OBP against SLG, adding text.

correspond to players with strong skills in getting on-base, but relatively weak in advancing runners home. In contrast the two points to the left of the major point cloud correspond to hitters who are better in slugging than in reaching base.

3.7   A Numeric Variable and a Factor Variable

When one collects a numeric variable such as OPS and a factor such as era, one is typically interested in comparing the distributions of the numeric variable across different values of the factor. In R, the `stripchart` function can be used to construct parallel stripcharts or number line graphs for values of the factor, and the `boxplot` constructs parallel boxplots (graphs of summaries of the numeric variable) across the factor.

Home run hitting has gone through dramatic changes in the history of baseball and suppose we are interested in exploring these changes over baseball eras. Suppose one focuses on the home run rate defined by HR / AB for our Hall of Fame players. We add a new variable HR.AB to the data frame hof:

```
hof$HR.Rate <- with(hof, HR / AB)
```

3.7.1   Parallel stripcharts

One constructs parallel stripcharts of HR.Rate by Era by using the stripchart function; the argument of the function has the form HR.Rate $\sim$ Era and we indicate by the data argument that these variables are part of the hof data frame.

```
stripchart(HR.Rate ~ Era, data=hof)
```

There are some problems with the basic display (not shown here). First, by default, the era labels are drawn parallel to the y axis and all of the labels are not displayed. Second, the points are drawn on top of each other and it is difficult to see all of the points. One can have the labels drawn perpendicular to the axes by using the argument option las=2. Unfortunately, when we do this, the labels don't fit in the space to the left of the plotting box. We can create more space by changing the location of the plot region. The location of the plot region is given by the graphical parameter plt which by default is the vector (0.09048276 0.95365517 0.17516995 0.85917710) which represent fractions (xlo, xhi, ylo, yhi) of the figure region. By assigning the xlo value the larger value of 0.2, we add more space to the left of the plot region. The par function is used to set graphical parameters; here it is used to change the plot region by setting the value of plt. The method="jitter" and pch=1 arguments in the stripchart function are used to jitter the points and make the points open circles.

```
par(plt=c(.2, .94, .145, .883))
stripchart(HR.Rate ~ Era, data=hof,
           method="jitter", pch=1, las=2)
```

This display in Figure 3.12 shows how the rate of hitting home runs has changed over eras. Home runs were rare in the 19th Century and Dead Ball eras. In the Lively Ball era, home run hitting was still relatively low, but there were some unusually good home run hitters such as Babe Ruth. The home run rates in the Integration, Expansion, and Free Agency eras were pretty similar.

3.7.2   Parallel boxplots

An alternative display for comparing distributions is the boxplot function. This function has the same basic argument HR.Rate $\sim$ Era. As before, we use the argument las=2 to indicate the labels are drawn perpendicular to the axes, horizontal=TRUE will display the boxplots horizontally and we add the label "HR Rate" to the x axis.

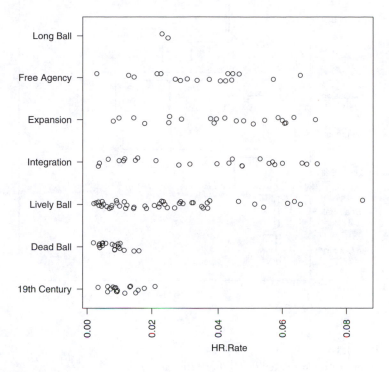

FIGURE 3.12
Stripcharts of home run rates of HOFers for each era.

```
par(plt=c(.2, .94, .145, .883))
boxplot(HR.Rate ~ Era, data=hof, las=2,
        horizontal=TRUE, xlab="HR Rate")
```

The parallel boxplot display is shown in Figure 3.13. Each rectangle in the display shows the location of the lower quartile, the median, and the upper quartile, and lines are drawn to the extreme values. Unusual points (outliers) that fall far from the rest of the distribution are indicated by open circle points. This graph confirms the observations we made when we viewed the stripchart display. Home run hitting was low in the first two eras and started to increase in the Lively Ball era. It is interesting that the only "outlier" among these Hall of Famers was Babe Ruth's career home run rate of 0.085.

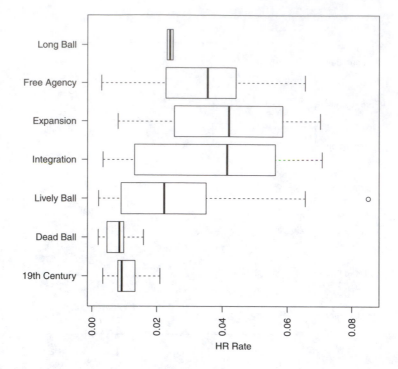

FIGURE 3.13
Boxplots of home run rates of HOFers for each era.

3.8   Comparing Ruth, Aaron, Bonds, and A-Rod

In Chapter 1, we constructed a graph comparing the career home run trajectories of four great sluggers in baseball history. In this section, we describe how we use R to create this graph. First, we need to read in the relevant data files into R. Next, we need to construct data frames containing the home run and age data for the sluggers. Last, we use R functions to construct the graph.

3.8.1   Getting the data

To obtain the graph, we need to collect the number of home runs, the at-bats, and the age for each season of each slugger's career. From the Lahman database, the relevant data files are `Master.csv` and `batting.csv`. From the master data file `Master.csv`, we obtain the player ids and birth years for the four players. The batting data file `batting.csv` is used to extract the home run and at-bats information.

We begin by reading in the Lahman master file, storing the file in the data frame `master`.

```
master <- read.csv("Master.csv")
```

From the Master table, we wish to extract the player id and the birth year for a particular player. Since we will be doing this operation for four players, it is convenient to write a new function `getinfo` to get this information for an arbitrary player of interest. The inputs to this function are the first and last names of the player and the output will be a list (a special data structure in R) giving the player's id and birth year. Some comments can be made about the R code in the function.

- The `subset` function is used to extract the row in the `master` data frame matching the player's first and last names; this row of this data frame is stored in the variable `playerline`.

- From the player information in the data frame `playerline`, we extract the player's birth year, birth month, birthday, and player id; these items are stored in the variables `birthyear`, `birthmonth`, `birthday`, and `name.code`.

- In Major League Baseball, a player's age for a season is defined to be his age on June 30. So we make a slight adjustment to a player's birth year depending if his birthday falls in the first six months or not. The adjusted birth year is stored in the variable `byear`. (The `ifelse` function is useful for assignments based on a condition; if `birthmonth <= 6` is TRUE, then `byear <- birthyear`, otherwise `byear <- birthyear + 1`.)

```
getinfo <- function(firstname, lastname){
  playerline <- subset(master,
        nameFirst==firstname & nameLast==lastname)
  name.code <- as.character(playerline$playerID)
  birthyear <- playerline$birthYear
  birthmonth <- playerline$birthMonth
  birthday <- playerline$birthDay
  byear <- ifelse(birthmonth <= 6,  birthyear, birthyear + 1)
  list(name.code=name.code, byear=byear)}
```

We use the function `getinfo` to get the information for the sluggers Babe Ruth, Hank Aaron, Barry Bonds, and Alex Rodriguez and store the information in variables. By displaying `ruth.info`, we see the player id and birth year for Babe Ruth.

```
ruth.info <- getinfo("Babe", "Ruth")
aaron.info <- getinfo("Hank", "Aaron")
bonds.info <- getinfo("Barry", "Bonds")
arod.info <- getinfo("Alex", "Rodriguez")
ruth.info
```

```
$name.code
[1] "ruthba01"

$byear
[1] 1895
```

Ruth's id code is "ruthba01" and he was born in 1895.

3.8.2   Creating the player data frames

Now that we have the player id codes and birth years, we use this information together with the Lahman batting file to create data frames for each player.

We read the Lahman batting data file "Batting.csv" into R, storing the file in the data frame `batting`.

```
batting <- read.csv("Batting.csv")
```

One of the variables in the batting data frame is `playerID`. To get the batting data for Babe Ruth, we use the `subset` function to extract the rows of the batting data from where `playerID` is equal to "ruthba01". We create a new variable `Age` defined to be the season year minus the player's birth year. (Recall that we made a slight modification to the `byear` variable so that one obtains a player's correct age for a season.)

```
ruth.data <- subset(batting, playerID == ruth.info$name.code)
ruth.data$Age <- ruth.data$yearID - ruth.info$byear
```

We perform similar commands to get batting data frames for the sluggers Hank Aaron, Barry Bonds, and Alex Rodriguez.

```
aaron.data <- subset(batting, playerID == aaron.info$name.code)
aaron.data$Age <- aaron.data$yearID - aaron.info$byear
bonds.data <- subset(batting, playerID == bonds.info$name.code)
bonds.data$Age <- bonds.data$yearID - bonds.info$byear
arod.data <- subset(batting, playerID == arod.info$name.code)
arod.data$Age <- arod.data$yearID - arod.info$byear
```

From the data frames `ruth.data`, `aaron.data`, `bonds.data`, and `arod.data`, it is straightforward to use R data analysis operations to compare the batting performance of these four sluggers.

3.8.3   Constructing the graph

We want to plot the cumulative home run counts for each of the four players against age. In each player data frame, the relevant variables are `HR` and `Age`. The function `cumsum` computes the cumulative sums of a vector. For example, below we illustrate the use of `cumsum` to compute the cumulative sums of the vector $\{1, 2, 3, 4\}$.

```
cumsum(c(1, 2, 3, 4))
[1]  1  3  6 10
```

We use the `plot` function to graph Ruth's cumulative home run count against his age. The arguments to `plot` indicate that a line graph will be drawn (`type="l"`) using a dotted line type (`lty=3`) of double-thickness (`lwd=2`). The `xlab` and `ylab` arguments label the horizontal and vertical axes and the `xlim` and `ylim` arguments give the limits of the two axes.

```
with(ruth.data, plot(Age, cumsum(HR), type="l", lty=3, lwd=2,
        xlab="Age", ylab="Career Home Runs",
        xlim=c(18, 45), ylim=c(0, 800)))
```

Using three applications of the `lines` function, three lines are added to the current graph corresponding to the cumulative home runs of Aaron, Bonds, and Rodriguez. Different line styles are applied by use of the `lty` argument so we can distinguish the four lines of the graph. Using the `legend` function, a legend is added to the graph connecting the line styles with the players. The argument to `legend` are the x and y coordinates of the location, a vector of character strings to display, the corresponding vector of line styles (`lty`), and the line width (`lwd`). Figure 3.14 displays the completed graph.

```
with(aaron.data, lines(Age, cumsum(HR), lty=2, lwd=2))
with(bonds.data, lines(Age, cumsum(HR), lty=1, lwd=2))
with(arod.data, lines(Age, cumsum(HR), lty=4, lwd=2))
legend(20, 700, legend=c("Bonds", "Aaron", "Ruth", "ARod"),
        lty=1 : 4, lwd=2)
```

3.9 The 1998 Home Run Race

The Retrosheet play-by-play files are helpful for learning about patterns of player performance during a particular baseball season. We illustrate the use of R to read in the files for the 1998 season and graphically view the famous home run duel between Mark McGwire and Sammy Sosa.

3.9.1 Getting the data

There are three important data files to read into R. One file denoted by "all1998.csv" contains all the play-by-play data for the 1998 season. The file "fields.csv" contains the names of all variables in the play-by-play file, and the data file "retrosheetIDS.csv" contains the player id codes used to extract data for particular players.

We begin by reading in the 1998 play-by-play data and storing it in the data frame `data1998`. The first line of this data file does not contain the

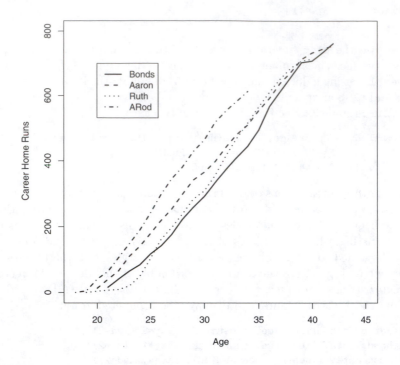

FIGURE 3.14
Career home runs by age for four great home run hitters in baseball history.

variable names, so we use the `header=FALSE` argument option. The header
information with the variable names is stored in the file "fields.csv". We read
this file, storing the names in the vector `fields`, and the `names` function is
used to add this header information to the data frame.

```
data1998 <- read.csv("all1998.csv", header=FALSE)
fields <- read.csv("fields.csv")
names(data1998) <- fields[, "Header"]
```

In the play-by-play database, the variable `BAT_ID` gives the identification
code for the player who is batting. To extract the batting data for McGwire
and Sosa, we need to find the codes for these two players. The data file "ret-
rosheetIDs.csv" is useful for this purpose. We read this data file into R and
store the information in the data frame `retro.ids`. There are three variables
in this data frame – `LAST`, `FIRST`, and `ID`. By use of the `subset` function, we
find the id code where `FIRST`-"Sammy" and `LAST`="Sosa". Likewise, we find
the id code corresponding to Mark McGwire; these codes are stored in the
variables `sosa.id` and `mac.id`.

```
retro.ids <- read.csv("retrosheetIDs.csv")
```

```
sosa.id <- as.character(subset(retro.ids,
          FIRST=="Sammy" & LAST=="Sosa")$ID)
mac.id <- as.character(subset(retro.ids,
          FIRST=="Mark" & LAST=="McGwire")$ID)
```

Now that the player id codes are obtained, McGwire's and Sosa's plate appearance data are extracted from the play-by-play data frame `data1998`. The two player data frames are stored in the variables `sosa.data` and `mac.data`.

```
sosa.data <- subset(data1998, BAT_ID == sosa.id)
mac.data <- subset(data1998, BAT_ID == mac.id)
```

3.9.2 Extracting the variables

For each player, we are interested in collecting the current number of home runs hit for each plate appearance and graphing the date against the home run count. For each player data frame, the two important variables are the date and the home run count. We write a new function `createdata` that will extract these two variables given a player's play-by-play batting data.

In the play-by-play data frame, the variable `GAME_ID` identifies the game location and date. For example, the value `GAME_ID` of "ARI199805110" indicates that this particular play occurred at the game played in Arizona on May 11, 1998. (The variable is displayed in the "location, year, month, day" format.) Using the `substr` function, we select the 4th through 11th characters of this string variable and assign this date to the variable `Date`. (The `as.Date` converts the date to the more readable "year-month-day" format.) Using the `order` function, we sort the play-by-play data from the beginning to the end of the season. The variable `EVENT_CD` contains the outcome of the batting play; a value `EVENT_CD` of 23 indicates that a home run has been hit. A new variable `HR` is defined to be either 1 or 0 depending if a home run occurred or not, and the new variable `cumHR` computes the cumulative number of home runs hit in the season. The output of the function is a new data frame containing the date and the cumulative number of home runs for all plate appearances during the season.

```
createdata <- function(d){
  d$Date <- as.Date(substr(d$GAME_ID, 4, 11),
                    format="%Y%m%d")
  d <- d[order(d$Date), ]
  d$HR <- ifelse(d$EVENT_CD == 23, 1, 0)
  d$cumHR <- cumsum(d$HR)
  d[, c("Date", "cumHR")]
}
```

We use the function `createdata` twice, once on Sosa's batting data and once on McGwire's batting data, obtaining the new data frames `mac.hr` and `sosa.hr`. We display the first few lines (using the `head` function) of `sosa.hr` to show the format of these new data frames.

```
mac.hr <- createdata(mac.data)
sosa.hr <- createdata(sosa.data)
head(sosa.hr)
            Date cumHR
71539 1998-03-31     0
71561 1998-03-31     0
71581 1998-03-31     0
71600 1998-03-31     0
71621 1998-03-31     0
71628 1998-04-01     0
```

3.9.3   Constructing the graph

Once these new data frames are created, it is straightforward to produce the graph of interest. The `plot` function constructs a graph of the cumulative home run count against the date. McGwire's data is plotted using a thick (`lwd=2`) line and Sosa's data is overlaid using the `lines` function and using a grey color (`col="grey"`). The `abline` function is used to add a horizontal line at the home run value of 62 and the `text` function is applied to place the text string "62" above this plotted line. We conclude using the `legend` function to identify McGwire and Sosa's home run trajectories. (See Figure 3.15.)

```
plot(mac.hr, type="l", lwd=2, ylab="Home Runs in the Season")
lines(sosa.hr, lwd=2, col="grey")
abline(h=62, lty=3)
text(10440, 65, "62")
legend(10440, 20, legend=c("McGwire (70)", "Sosa (66)"),
       lwd=2, col=c("black", "grey"))
```

3.10   Further Reading

A good reference to the traditional graphics system in R is Murrell (2011). Kabacoff (2011) together with the Quick-R website at `www.statmethods.net` provide a useful reference for specific graphics functions. Chapter 4 of Albert and Rizzo (2012) provides a number of examples of modifying traditional graphics in R such as changing the plot type and symbol, using color, and overlying curves and mathematical expressions.

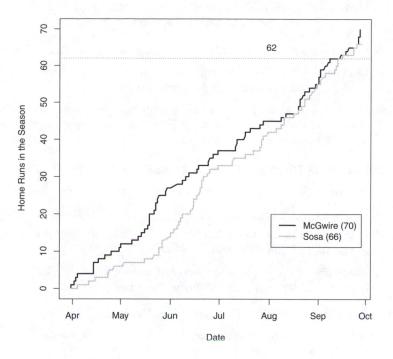

FIGURE 3.15
Seasonal home runs for Mark McGwire and Sammy Sosa during the 1998 race.

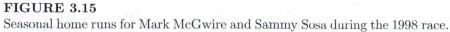

3.11   Exercises

1. (**Hall of Fame Pitching Dataset**)

   The data file "hofpitching.csv" contains the career pitching statistics for all of the pitchers inducted in the Hall of Fame. This data file can be read into R by means of the `read.csv` function.

   ```
   hofpitching <- read.csv("hofpitching.csv")
   ```

   The variable BF is the number of batters faced by a pitcher in his career. Suppose we group the pitchers by this variable using the intervals (0, 10,000), (10,000, 15,000), (15,000, 20,000), (20,000, 30,000). One can reexpress the variable BF to the grouped variable `BF.group` by use of the `cut` function.

   ```
   hofpitching$BF.group <- with(hofpitching,
          cut(BF, c(0, 10000, 15000, 20000, 30000),
   ```

```
labels=c("Less than 10000", "(10000, 15000)",
         "(15000, 20000)", "more than 20000")))
```

(a) Construct a frequency table of `BF.group` using the `table` function.

(b) Construct a bar graph of the output from `table`. How many HOF pitchers faced more than 20,000 pitchers in their career?

(c) Construct a pie graph of the `BF.group` variable. Compare the effectiveness of the bar graph and pie graph in comparing the frequencies in the four intervals.

2. (**Hall of Fame Pitching Dataset (Continued)**)

The variable `WAR` is the total wins above replacement of the pitcher during his career.

(a) Using the `hist` function, construct a histogram of `WAR` for the pitchers in the Hall of Fame dataset.

(b) There are two pitchers who stand out among all of the Hall of Famers on the total `WAR` variable. Identify these two pitchers.

3. (**Hall of Fame Pitching Dataset (Continued)**)

To understand a pitcher's season contribution, suppose we define the new variable `WAR.Season` defined by

```
hofpitching$WAR.Season <- with(hofpitching, WAR / Yrs)
```

(a) Use the `stripchart` function to construct parallel stripcharts of `WAR.Season` for the different levels of `BP.group`.

(b) Use the `boxplot` function to construct parallel boxplots of `WAR.Season` across `BP.group`.

(c) Based on your graphs, how does the wins above replacement per season depend on the number of batters faced?

4. (**Hall of Fame Pitching Dataset (Continued)**)

Suppose we limit our exploration to pitchers whose mid-career was 1960 or later. We first define the `MidYear` variable and then use the `subset` function to construct a data frame consisting of only these 1960+ pitchers.

```
hofpitching$MidYear <- with(hofpitching, (From + To) / 2)
hofpitching.recent <- subset(hofpitching, MidYear >= 1960)
```

(a) By use of the `order` function, order the rows of the data frame by the value of `WAR.Season`.

(b) Construct a dot plot of the values of `WAR.Season` where the labels are the pitcher names.

(c) Which two 1960+ pitchers stand out with respect to wins above replacement per season?

5. (**Hall of Fame Pitching Dataset (Continued)**)
The variables `MidYear` and `WAR.Season` are defined in the previous exercises.

   (a) Construct a scatterplot of `MidYear` (horizontal) against `WAR.Season` (vertical).

   (b) Is there a general pattern in this scatterplot? Explain.

   (c) There are two pitchers whose mid careers were in the 1800s who had relatively low `WAR.Season` values. Use the `identify` function with the scatterplot to find the names of these two pitchers.

6. (**Working with the Lahman Batting Dataset**)

   (a) Read the Lahman "Master.csv" and "batting.csv" data files into R.

   (b) Use the `getinfo` to obtain three data frames for the season batting statistics for the great hitters Ty Cobb, Ted Williams, and Pete Rose.

   (c) Add the variable `Age` to each data frame corresponding to the ages of the three players.

   (d) Using the `plot` function, construct a line graph of the cumulative hit totals against age for Pete Rose.

   (e) Using the `lines` function, overlay the cumulative hit totals for Cobb and Williams.

   (f) Write a short paragraph summarizing what you have learned about the hitting pattern of these three players.

7. (**Working with the Retrosheet Play-by-Play Dataset**)
In Section 3.9, we used the Retrosheet play-by-play data to explore the home run race between Mark McGwire and Sammy Sosa in the 1998 season. Another way to compare the patterns of home run hitting of the two players is to compute the spacings, the number of plate appearances between home runs.

   (a) Following the work in Section 3.9, create the two data frames `mac.data` and `sosa.data` containing the batting data for the two players.

   (b) Use the following R commands to restrict the two data frames to the plays where a batting event occurred. (The relevant variable `BAT_EVENT_FL` is either `TRUE` or `FALSE`.)

```
mac.data <- subset(mac.data, BAT_EVENT_FL == TRUE)
sosa.data <- subset(sosa.data, BAT_EVENT_FL == TRUE)
```

(c) For each data frame, create a new variable `PA` that numbers the plate appearances 1, 2, ... (The function `nrow` gives the number of rows of a data frame.)

```
mac.data$PA <- 1:nrow(mac.data)
sosa.data$PA <- 1:nrow(sosa.data)
```

(d) The following commands will return the numbers of the plate appearances when the players hit home runs.

```
mac.HR.PA <- mac.data$PA[mac.data$EVENT_CD==23]
sosa.HR.PA <- sosa.data$PA[sosa.data$EVENT_CD==23]
```

(e) Using the R function `diff`, the following commands compute the spacings between the occurrences of home runs.

```
mac.spacings <- diff(c(0, mac.HR.PA))
sosa.spacings <- diff(c(0, sosa.HR.PA))
```

(f) By use of the `summary` and `hist` functions on the vectors `mac.spacings` and `sosa.spacings`, compare the home run spacings of the two players.

4

The Relation Between Runs and Wins

CONTENTS

4.1	Introduction	87
4.2	The Teams Table in Lahman's Database	88
4.3	Linear Regression	89
4.4	The Pythagorean Formula for Winning Percentage	93
4.5	The Exponent in the Pythagorean Formula	95
4.6	Good and Bad Predictions by the Pythagorean Formula	96
4.7	How Many Runs for a Win?	99
4.8	Further Reading	102
4.9	Exercises	102

4.1 Introduction

The goal of a baseball team is, just like in any other sport, winning games. Similarly, the goal of the baseball analyst is being able to measure what happens on the field in term of wins. Answering a question such as "Who is the better player between Brett Gardner and Prince Fielder?" becomes an easier task if one succeeds in estimating how much Gardner's speed and slick fielding contribute to his team's victories and how many wins can be attributed to Prince's powerful bat.

Victories are obtained by outscoring opponents, thus the percentage of wins obtained by a team over the course of a season is strongly related with the number of runs it scores and allows. This chapter explores the relationship between runs and wins. Understanding this relationship is a critical step towards answering questions on players' value. In fact, while it's impossible to directly quantify the impact of players in terms of wins, it will be seen in the following chapters that it is possible to measure their contributions in term of runs.

4.2   The Teams Table in Lahman's Database

The `teams.csv` file from Lahman's database contains seasonal stats for major league teams going back to the first professional season in 1871. We begin by loading this database file into R and exploring its contents by looking at the final lines of this dataset, using the `tail` function.

```
teams <- read.csv("teams.csv")
tail(teams)
```

	yearID	lgID	teamID	franchID	divID	Rank	G	Ghome	W	L
2683	2011	NL	LAN	LAD	W	3	161	NA	82	79
2684	2011	NL	COL	COL	W	4	162	NA	73	89
2685	2011	NL	SDN	SDP	W	5	162	NA	71	91

	DivWin	WCWin	LgWin	WSWin	R	AB	H	X2B	X3B	HR	BB	SO
2683	N	N	N	N	644	5436	1395	237	28	117	498	1087
2684	N	N	N	N	735	5544	1429	274	40	163	555	1201
2685	N	N	N	N	593	5417	1284	247	42	91	501	1320

	SB	CS	HBP	SF	RA	ER	ERA	CG	SHO	SV	IPouts	HA	HRA	BBA
2683	126	40	45	43	612	563	3.56	7	17	40	4296	1287	132	507
2684	118	42	57	44	774	713	4.44	5	7	41	4343	1471	176	522
2685	170	44	48	47	611	551	3.43	0	10	44	4348	1324	125	521

	SOA	E	DP	FP	name	park
2683	1265	85	121	0.986	Los Angeles Dodgers	Dodger Stadium
2684	1118	98	156	0.984	Colorado Rockies	Coors Field
2685	1139	94	138	0.985	San Diego Padres	Petco Park

	attendance	BPF	PPF	teamIDBR	teamIDlahman45	teamIDretro
2683	2935139	98	98	LAD	LAN	LAN
2684	2909777	116	116	COL	COL	COL
2685	2143018	92	92	SDP	SDN	SDN

The description of every column is provided in the `readme` file accompanying the Lahman's database.

Suppose that one is interested in relating the proportion of wins with the runs scored and runs allowed for all of the teams. Towards this goal, the relevant fields of interest in this table are the number of games played `G`, the number of team wins `W`, the number of losses `L`, the total number of runs scored `R`, and the total number of runs allowed `RA`. A new data frame `myteams` is created containing only the above five columns plus information on the team (`teamID`), the season (`yearID`), and the league (`lgID`). We are interested in studying the relationship between wins and runs for recent seasons, so by use of the `subset` function, we focus our exploration to seasons since 2001.

```
myteams <- subset(teams, yearID > 2000)[ , c("teamID", "yearID",
  "lgID", "G", "W", "L", "R", "RA")]
```

```
tail(myteams)
```

```
      teamID yearID lgID   G  W  L   R  RA
2680     FLO   2011   NL 162 72 90 625 702
2681     ARI   2011   NL 162 94 68 731 662
2682     SFN   2011   NL 162 86 76 570 578
2683     LAN   2011   NL 161 82 79 644 612
2684     COL   2011   NL 162 73 89 735 774
2685     SDN   2011   NL 162 71 91 593 611
```

The run differential is defined as the difference between the runs scored and the runs allowed by a team. The winning proportion is the fraction of games won by a team. In baseball (and generally in sports) *winning percentage* is commonly used instead of the more appropriate *winning proportion*. In the remainder of this chapter we have chosen to adopt the most widely used term. Two new variables RD (run differential) and Wpct (winning percentage) are calculated with the following lines of code.

```
myteams$RD <- with(myteams, R - RA)
myteams$Wpct <- with(myteams, W / (W + L))
```

A scatterplot of the run differential and the winning percentage using the `plot` function gives a first indication on the association between the two variables.

```
plot(myteams$RD, myteams$Wpct,
     xlab="run differential",
     ylab="winning percentage")
```

As expected, Figure 4.1 shows a strong positive relationship – teams with large run differentials are more likely to be winning.

4.3 Linear Regression

One simple way to predict a team's winning percentage using runs scored and allowed is with linear regression. A simple linear model is

$$Wpct = a + b \times RD + \epsilon,$$

where a and b are unknown constants and ϵ is the error term which captures all other factors influencing the dependent variable (Wpct). This is a special case of a linear model fit using the `lm` function from the `stats` package (which is installed and loaded in R by default). The most basic call to the function requires a formula, specified as `response ~ predictor1 + predictor2 + ...`, `data=dataset`, in which

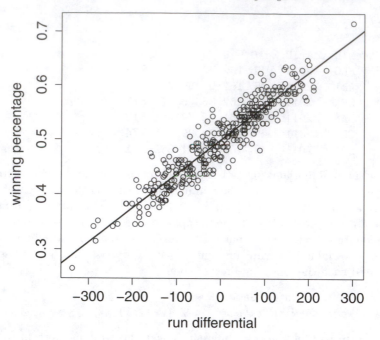

FIGURE 4.1
Scatterplot of team run differential against team winning percentage for major
league teams from 2001 to 2011. A best-fitting line is overlaid on top of the
scatterplot.

the variable to be predicted (the *dependent variable*) is indicated on the left
side of the tilde character and the the variables used to predict the response
are specified on the right side. In the following illustration of the `lm` function,
the `data` argument in `lm` is used to specify which data frame to use.

```
linfit <- lm(Wpct ~ RD, data=myteams)

linfit
Call:
lm(formula = Wpct ~ RD, data=myteams)

Coefficients:
(Intercept)              RD
   0.499992        0.000623
```

The fitted line in the plot of Figure 4.1 is obtained with the `abline` function[1] and the `coef` function for extracting the model (`linfit`) coefficients.

```
abline(a=coef(linfit)[1], b=coef(linfit)[2], lwd=2)
```

From the above output, a team's winning percentage can be estimated from its run differential RD by the equation:

$$Wpct = 0.499992 + 0.000623 \times RD$$

This formula tells us that a team with a run differential of zero ($RD = 0$) will win half of its games (estimated intercept $\approx .500$) which is reasonable. In addition, a one-unit increase in run differential corresponds to an increase of 0.000623 in winning percentage. To give further insight into this relationship, a team scoring 750 runs and allowing 750 runs is predicted to win half of its games corresponding to 81 games in a typical MLB season of 162 games. In contrast, a team scoring 760 runs and allowing 750 has a run differential of $+10$ and is predicted to have a winning percentage of $0.500 + 10 \cdot 0.000623 \approx 0.506$. A winning percentage of 0.506 in a 162-game schedule corresponds to 82 wins. Thus an increase of 10 runs in the run differential of a team corresponds, according to the straight-line model, to an additional win in the standings.

One concern is that predictions from this fitted line can assume values outside the range $[0, 1]$. For example, a hypothetical team that outscores its opponent by a total of 805 runs would be predicted to win more than 100 percent of its games which is impossible. However, since over 99 percent of teams throughout major league baseball history have run differentials between -350 and +350, the straight-line model is a reasonable approximation.

Once one has a fitted model, the function `predict` can be used to calculate the predicted values from the model, while the function `residuals` computes the difference between the response values and the fitted values (i.e., between the actual and the estimated winning percentages).

```
myteams$linWpct <- predict(linfit)
myteams$linResiduals <- residuals(linfit)
```

Figure 4.2 displays a plot of the residuals against the run differential using the following code. The `abline` and `text` functions have been introduced in Chapter 3, while `points` is used to draw data points at specified coordinates (in particular, the `points` and `text` functions are used to mark and label a few anomalous data points).

```
plot(myteams$RD, myteams$linResiduals,
     xlab="run differential",
     ylab="residual")
```

---

[1]In this case, rather than a single argument `h` or `v` for plotting an horizontal or a vertical line, `abline` is supplied with two arguments `a` and `b` indicating the intercept and the slope of the line to be drawn.

```
abline(h=0, lty=3)
points(c(68, 88), c(.0749, -.0733), pch=19)
text(68, .0749, "LAA '08", pos=4, cex=.8)
text(88, -.0733, "CLE '06", pos=4, cex=.8)
```

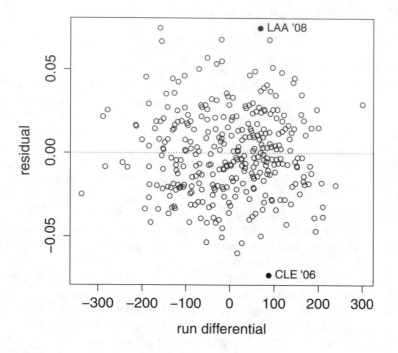

FIGURE 4.2
Residuals versus run differential for the fitted linear model. Two large residuals are labeled corresponding to the 2008 Los Angeles Angels and the 2006 Cleveland Indians.

Residuals can be interpreted as the error of the linear model in predicting the actual winning percentage. Thus the points in Figure 4.2 farthest from the zero line correspond to the teams where the linear model fared worst in predicting the winning percentage.

One of the extreme values at the top of the residual graph in Figure 4.2 corresponds to the 2008 Los Angeles Angels: given their +68 run differential, they were supposed, according to the linear equation (4.3), to have a 0.542 winning percentage; they ended the season at 0.617. The residual value for this team is $0.617 - 0.542 = 0.075$. At the other end of the spectrum, the 2006 Cleveland Indians, with a +88 run differential, are seen as a 0.555 team by the linear model, but they actually finished at a mere 0.481, corresponding to the residual $0.481 - 0.555 = -0.073$.

The average value of the residuals for this model is equal to zero, which means that the model predictions are equally likely to overestimate than to underestimate the winning percentage, or that the method for fitting the model is unbiased. In order to estimate the average magnitude of the errors, one first squares the residuals so that each error has a positive value, calculates the mean of the squared residuals, and takes the square root of such mean value to get back to the original scale. The value so calculated is the root mean square error, abbreviated as RMSE. (The square root function `sqrt` is introduced.)

```
mean(myteams$linResiduals)
[1] -2.952603e-19
linRMSE <- sqrt(mean(myteams$linResiduals ^ 2))
linRMSE
[1] 0.02507176
```

Approximately two thirds of the residuals fall between $-RMSE$ and $+RMSE$, while 95% of the residuals are between $-2 \cdot RMSE$ and $2 \cdot RMSE$.[2] These statements can be confirmed with the following lines of code. (The function `abs` computes the absolute value.)

```
nrow(subset(myteams, abs(linResiduals) < linRMSE)) /
  nrow(myteams)
[1] 0.6757576

nrow(subset(myteams, abs(linResiduals) < 2 * linRMSE)) /
  nrow(myteams)
[1] 0.9545455
```

Above the function `nrow` is used to obtain the number of rows of a data frame. In the numerators of the expressions, we obtain the number of residuals sizes (computed using the `abs` function) that are smaller than one and two $RMSE$. The computed fractions are close to the 68% and 95% values stated above.

4.4   The Pythagorean Formula for Winning Percentage

Bill James, regarded as the father of Sabermetrics, empirically derived the following non-linear formula to estimate winning percentage, called the Pythagorean expectation.

$$Wpct = \frac{R^2}{R^2 + RA^2}$$

---

[2]Equivalently, it can be stated that, over a 162-game season, the number of wins predicted by the linear model comes within four wins of the actual number of wins in two-thirds of the cases, while for 19 out of 20 teams the difference is not higher than 8 wins.

One can use this formula to predict winning percentages by use of the following R code.

```
myteams$pytWpct <- with(myteams, R ^ 2 / (R ^ 2 + RA ^ 2))
```

Here the residuals need to be calculated explicitly, but that's not a hard task. A new variable `pytResiduals` is defined that is the difference between the actual and predicted winning percentages. The RMSE is computed for these new predictions.

```
myteams$pytResiduals <- myteams$Wpct - myteams$pytWpct
sqrt(mean(myteams$pytResiduals ^ 2))
[1] 0.02545247
```

The RMSE calculated on the Pythagorean predictions is similar in value to the one calculated with the linear predictions (it's actually higher for the 2000-2011 data we have been using here). Thus it does not seem justifiable using a more complex model. However, the Pythagorean expectation has several desirable properties missing in the linear model. Both of these advantages can be illustrated with several examples.

Suppose there exists a powerhouse team that scores an average of ten runs per game, while allowing a close to average five runs per game. In a 162-game schedule, this team would score 1620 runs, while allowing 810, for a run differential of 810. Replacing RD with 810 in the linear equation, one obtains a winning percentage of over 1, which is impossible. On the other hand, replacing R and RA with 1620 and 810 respectively in the Pythagorean expectation, the resulting winning percentage is equal to 0.8, a more reasonable prediction. A second hypothetical team has pitchers who never allow runs, while the hitters always manage to score the only run they need. Such a team will score 162 runs in a season and win all of its games, but the linear equation would predict it to be merely a .601 team. The Pythagorean formula, instead, correctly predicts this team to win all of its games.

While neither of the above examples is ever going to materialize, there are some extreme situations in modern baseball history. For example, the 2001 Seattle Mariners had 116 wins and 46 losses for a +300 run differential and the 2003 Detroit Tigers had a 43-119 recored with a -337 run differential. In these unlikely scenarios, the Pythagorean formula will give more sensible winning percentage estimates.

Recall our statement at the end of the introductory section that the runs-to-wins relationship is crucial in assessing the contribution of players to their team's wins. Once we estimate the number of runs players contribute to their teams (as it will be shown in the following chapters), runs-to-wins formulas can be used to convert these run values to wins. One can now answer questions like "Home many wins would a lineup of nine Albert Pujols' accumulate in a season?" For these kind of investigations, the scenarios in which the linear formula break down are more likely to occur, thus highlighting the need for a formula such as the Pythagorean expectation that gives reasonable predictions.

4.5 The Exponent in the Pythagorean Formula

Subsequent refinements to the Pythagorean formula by Bill James and other analysts have aimed at finding an exponent which would give a better fit relative to the originally proposed exponent value of 2. In this section, we describe how one finds the value of the Pythagorean exponent leading to predictions closest to the actual winning percentages.

Replacing the value 2 in the exponents in the Pythogorean expectation with an unknown variable k, the formula is written as

$$W\% = \frac{R^k}{R^k + RA^k}$$

With some algebra, the equation can be rewritten as follows:

$$\frac{W}{L} = \frac{R^k}{RA^k}$$

Taking the logarithm on both sides of the equation (using the function `log`), one obtains the linear relationship

$$log\left(\frac{W}{L}\right) = k \cdot log\left(\frac{R}{RA}\right)$$

The value of k can now be estimated using linear regression, where the response variable is `log(W/L)` and the predictor is `log(R/RA)`. In the following R code, we compute the logarithm of the ratio of wins to losses, the logarithm of the ratio of runs to runs allowed, and fit a simple linear model with these transformed variables. (In the call of the `lm` function, a model with a zero intercept is indicated by a zero term on the right side of the formula.)

```
myteams$logWratio <- log(myteams$W / myteams$L)
myteams$logRratio <- log(myteams$R / myteams$RA)
pytFit <- lm(logWratio ~ 0 + logRratio, data=myteams)
pytFit

Call:
lm(formula = logWratio ~ 0 + logRratio, data=myteams)

Coefficients:
logRratio
     1.903
```

The R output suggests a Pythagorean exponent of 1.903 which is significantly smaller than the value 2.

4.6   Good and Bad Predictions by the Pythagorean Formula

The 2011 Boston Red Sox scored 875 runs, while allowing 737. According to the Pythagorean formula with exponent 2, they were expected to win 95 games – we obtain this number by plugging 875 and 737 into the Pythagorean formula and multiplying by the number of games in a season:

$$162 \times \frac{875^2}{875^2 + 737^2} \approx 95$$

The Red Sox actually won 90 games. The five games difference was quite costly to the Red Sox, as they missed clinching the Wild Card (that went to the Rays) in the final game (actually in the final minute) of the season. The Pythagorean formula is more on target with the Tampa Bay Rays of the same season, as the prediction of 92 (coming from their 707 runs scored versus 614 runs allowed) is just a bit higher than the actual 91.

Why does the Pythagorean formula miss so poorly on the Red Sox? In other words, why did they win five less games than expected from their run differential? Let's have a look at their season game by game.

The `gl2011.txt` (a game log file downloaded from Retrosheet, see Section 1.3.3) contains detailed information on every game played in the 2011 season. The following commands load the file into R, select the lines pertaining to the Red Sox games, and keep only the runs related columns.

```
gl2011 <- read.table("gl2011.txt", sep=",")
glheaders <- read.csv("retrosheet/game_log_header.csv")
names(gl2011) <- names(glheaders)
BOS2011 <- subset(gl2011, HomeTeam=="BOS" | VisitingTeam=="BOS")[
  , c("VisitingTeam", "HomeTeam", "VisitorRunsScored",
      "HomeRunsScore")]
head(BOS2011)
```

	VisitingTeam	HomeTeam	VisitorRunsScored	HomeRunsScore
16	BOS	TEX	5	9
31	BOS	TEX	5	12
45	BOS	TEX	1	5
61	BOS	CLE	1	3
76	BOS	CLE	4	8
88	BOS	CLE	0	1

Using the results of every game featuring the Boston team, run differentials (`ScoreDiff`) are calculated both for games won and lost and a column `W` is added indicating whether the Red Sox won the game.

```
BOS2011$ScoreDiff <- with(BOS2011, ifelse(HomeTeam == "BOS",
  HomeRunsScore - VisitorRunsScored,
  VisitorRunsScored - HomeRunsScore))
BOS2011$W <- BOS2011$ScoreDiff > 0
```

Summary statistics are computed on the run differentials for games won and for games lost using the `aggregate` function. For this function, the first argument is the variable for which the summary statistics are to be calculated (the absolute value of the score differential), the second argument is a list of grouping factors (in this case only one factor is specified–whether the game resulted in a win for Boston), and the final argument is the summarizing function we want to apply (`summary` which was introduced in Chapter 2).

```
aggregate(abs(BOS2011$ScoreDiff), list(W=BOS2011$W), summary)
```

	W	x.Min.	x.1st Qu.	x.Median	x.Mean	x.3rd Qu.	x.Max.
1	FALSE	1.000	1.000	3.000	3.458	4.000	11.000
2	TRUE	1.000	2.000	4.000	4.300	6.000	14.000

The 2011 Red Sox had their victories decided by a larger margin than their losses (4.3 vs 3.5 runs on average), leading to their underperformance of the Pythagorean prediction by five games. A team overperforming (or underperforming) its Pythagorean winning percentage is often seen, in sabermetrics circles, as being lucky (or unlucky), and consequently is expected to get closer to its expected line as the season progresses.

A team can overperform its Pythagorean winning percentage by winning a disproportionate number of close games. This claim can be confirmed by a brief data exploration. With the following lines of code, a data frame (`results`) is created from the previously loaded 2011 game logs, containing the names of the teams and the runs scored. Two new columns are created, the variable `winner` contains the abbreviation of the winning team and a second variable `diff` contains the margin of victory.

```
results <- gl2011[,c("VisitingTeam", "HomeTeam",
  "VisitorRunsScored", "HomeRunsScore")]
results$winner <- ifelse(results$HomeRunsScore >
  results$VisitorRunsScored, as.character(results$HomeTeam),
  as.character(results$VisitingTeam))
results$diff <- abs(results$VisitorRunsScored -
  results$HomeRunsScore)
```

Suppose we focus on the games won by only one run. The data frame `onerungames` is created containing only the games decided by one run, and the `table` function is used to count the number of wins in such contests for each team. The `as.data.frame` function converts this table to a data frame and the `names` function adds names to the columns.

```
onerungames <- subset(results, diff == 1)
onerunwins <- as.data.frame(table(onerungames$winner))
names(onerunwins) <- c("teamID", "onerunW")
```

Using the `myteams` data frame previously created, we look at the relation between the Pythagorean residuals and the number of one-run victories. Note that the team abbreviation for the Angels needs to be changed because it is coded as "LAA" in the Lahman's database and as "ANA" in the Retrosheet game logs.

```
teams2011 <- subset(myteams, yearID == 2011)
teams2011[teams2011$teamID == "LAA", "teamID"] <- "ANA"
teams2011 <- merge(teams2011, onerunwins)
plot(teams2011$onerunW, teams2011$pytResiduals,
     xlab="one run wins",
     ylab="Pythagorean residuals")
```

The final line of code produces the plot in Figure 4.3 which shows a positive relationship between the number of one-run games won and the Pythagorean residuals. The `identify` function is used to identify teams on the plot.

```
identify(teams2011$onerunW, teams2011$pytResiduals,
  labels=teams2011$teamID)
```

Figure 4.3 shows that San Francisco is a team with a large number of one-run victories and a large positive Pythogorean residual. In contrast, San Diego had few one-run victories and a negative residual.

Winning a disproportionate number of close games is sometimes attributed to plain luck. However teams with certain attributes may be more likely to systematically win contests decided by a narrow margin. For example, teams with top quality closers will tend to preserve small leads, and will be able to overperform their Pythogorean expected winning percentage. To check this conjecture, we look at the data.

The `pitching.csv` file in the Lahman's database contains individual seasonal pitching stats. This file is read into R and the `subset` function is used select the pitchers/seasons where more than 50 games were finished with an ERA lower than 2.50. The data frame `top_closers` contains only the columns identifying the pitcher, the season and the team.

```
pit <- read.csv("pitching.csv")
top_closers <- subset(pit, GF > 50 & ERA < 2.5)[ ,c("playerID",
  "yearID", "teamID")]
```

The `top_closers` data frame is merged with our `myteams` dataset, creating the new `teams_top_closers` data frame – this contains the teams featuring a top closer. Summary statistics on the Pythagorean residuals are obtained using the `summary` function.

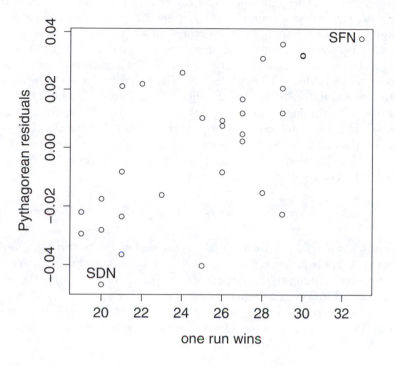

FIGURE 4.3

Scatterplot of number of one-run games won and Pythagorean residuals for major league teams in 2011.

```
teams_top_closers <- merge(myteams, top_closers)
summary(teams_top_closers$pytResiduals)
    Min.   1st Qu.    Median      Mean   3rd Qu.      Max.
-0.048690 -0.011660  0.003359  0.005189  0.022990  0.071400
```

The mean of the residuals is only slightly above zero (0.005189), but when one multiplies it by the number of games in a season (162), one finds that teams with a top closer win, on average, 0.8 games more than would be predicted by the Pythagorean formula.

4.7   How Many Runs for a Win?

Readers familiar with websites like www.insidethebook.com, www.hardballtimes.com, and www.baseballprospectus.com/ are surely familiar with the ten-runs-equal-a-win rule of thumb. Over the course of a season, a team scoring ten more runs is likely to have one more win in the standings. The number comes directly from the Pythagorean formula with an exponent of two. Suppose a team scores an average of five runs per game, while allowing the same number of runs. In a 162-game season, the team would score (and allow) 810 runs. Inserting 810 in the Pythagorean formula one gets (as expected) a perfect .500 expected winning percentage with 81 wins. If one substitutes 810 with 820 for the number of runs scored in the formula, one obtains a .506 winning percentage that translates to 82 wins in 162 games. The same result is obtained for a team scoring 810 runs and allowing 800.

Ralph Caola has derived the number of extra runs needed to get an extra win in a more rigorous way using calculus. He starts from the equivalent representation of the Pythagorean formula.

$$W = G \cdot \frac{R^2}{R^2 + RA^2}$$

If one takes a partial derivative of the right side of the above equation with respect to R, holds RA constant, the result is the incremental number of wins per run scored. Taking the reciprocal of this result, one can derive the number of runs needed for an extra win.

R is capable of calculating partial derivatives, thus we can retrace Ralph's steps in R by using the functions D and `expression` to take the partial derivative of $R^2/(R^2 + RA^2)$ with respect to R.

```
D(expression(G * R ^ 2 / (R ^ 2 + RA ^ 2)), "R")
G * (2 * R)/(R^2 + RA^2)  G * R^2 * (2 * R)/(R^2 + RA^2)^2
```

Unfortunately R does not do the simplifying. The reader has the choice of either doing the tedious work himself or believing the final equation for incremental runs per win (`IR/W`) is the following[3]:

$$IR/W = \frac{\left(R^2 + RA^2\right)^2}{2 \cdot G \cdot R \cdot RA^2}$$

If R and RA are expressed in runs per game, the G is removed from the above formula.

Using this formula, one can compute the incremental runs needed per one win for various runs scored/runs allowed scenarios. As a first step, a function IR is created to calculate the incremental runs, according to Caola's formula; this function takes runs scored per game and runs allowed per game as arguments.

```
IR <- function(RS=5, RA=5){
 round((RS ^ 2 + RA ^ 2)^2 / (2 * RS * RA ^ 2), 1)
}
```

---

[3]The formula is the result of algebraic simplification and taking the reciprocal.

This function is used to create a table for various runs scored/runs allowed combinations. We perform this step by using the functions `seq` and `expand.grid`. The `seq` function is used create a vector containing a regular sequence specifying, as arguments, the start value, the end value, and the increment value. Here `seq` creates a vector of values from 3 to 6 in increments of 0.5. Then the `expand.grid` function is used to obtain a data frame containing all the combinations of the elements of the supplied vectors. In the following code the first and the final few lines of the new data frame `IRtable` are displayed.

```
IRtable <- expand.grid(RS=seq(3, 6, .5), RA=seq(3, 6, .5))
rbind(head(IRtable), tail(IRtable))
     RS RA
1   3.0  3
2   3.5  3
3   4.0  3
4   4.5  3
5   5.0  3
6   5.5  3
44  3.5  6
45  4.0  6
46  4.5  6
47  5.0  6
48  5.5  6
49  6.0  6
```

Finally, the incremental runs are calculated for the various scenarios. The `xtabs` function in the second line of the following code is used to show the results in a tabular form. The formula specified as the first argument has the variable which populates the cells on the left side of the tilde character, while on the right side the cross-classifying variables are separated by a + sign.

```
IRtable$IRW <- IR(IRtable$RS, IRtable$RA)
xtabs(IRW ~ RS + RA, data=IRtable)
     RA
RS        3  3.5    4  4.5    5  5.5    6
  3     6.0  6.1  6.5  7.0  7.7  8.5  9.4
  3.5   7.2  7.0  7.1  7.5  7.9  8.5  9.2
  4     8.7  8.1  8.0  8.1  8.4  8.8  9.4
  4.5  10.6  9.6  9.1  9.0  9.1  9.4  9.8
  5    12.8 11.3 10.5 10.1 10.0 10.1 10.3
  5.5  15.6 13.4 12.2 11.4 11.1 11.0 11.1
  6    18.8 15.8 14.1 13.0 12.4 12.1 12.0
```

Looking at the results we notice that the rule of ten is appropriate in typical run scoring environments (4 to 5 runs per game). However, in very low scoring environments (the upper-left corner of the table), a lower number of runs is needed to gain an extra win; on the other hand, in high scoring environments (lower-right corner), one needs a larger number of runs for an added win.

4.8   Further Reading

Bill James first mentioned his Pythagorean formula in James (1980) which, like other early works by James, was self-published and it is currently hard to find. Reference to the formula is present in James (1982), the first edition published by Ballantine Books. Davenport and Woolner (1999) and Heipp (2003) revisited Bill James' formula, deriving exponents that vary according to the total runs scored per game. Caola (2003) algebraically derived the relation between run scored and allowed and winning percentage. Star (2011) recounts the final moments of the 2011 regular season, when in the turn of a few minutes the Rays and the Red Sox fates dramatically turned ; the page also features a twelve-minute video chronicling the events of the wild September 28, 2011 night.

4.9   Exercises

1. (**Relationship Between Winning Percentage and Run Differential Across Decades**)

   Section 4.3 used a simple linear model to predict a team's winning percentage based on its run differential. This model was fit using team data since the 2001 season.

   (a) Refit this linear model using data from the seasons 1961-1970, the seasons 1971-1980, the seasons 1981-1990, and the seasons 1991-2000.

   (b) Compare across the five decades the predicted winning percentage for a team with a run differential of 10 runs.

2. (**Pythagorean Residuals for Poor and Great Teams in the 19th Century**)

   As baseball was evolving into its ultimate form, nineteenth century leagues often featured abysmal teams that did not even succeed in finishing their season, as well as some dominant clubs.

   (a) Fit a Pythagorean formula model to the run-differential, win-loss data for teams who played in the 19th century.

   (b) By inspecting the residual plot of your fitted model from (a), did the great and poor teams in the 19th century do better or worse than one would expect on the basis of their run differentials?

3. (**Exploring the Manager Effect in Baseball**)

Retrosheet game logs report, for every game played, the managers of both teams.

    (a) Select a period of your choice (encompassing at least ten years) and fit the Pythagorean formula model to the run-differential, win-loss data.

    (b) On the basis of your fit in part (a) and the list of managers, compile a list of the managers who most overperformed their Pythagorean winning percentage and the managers who most underperformed it.

4. (**Pythagorean Relationship for Other Sports**)

Bill James' Pythagorean formula has been used for predicting winning percentage in other sports. Since the pattern of scoring is very different among sports (compare for example points in basketball and goals in soccer), the formula needs to be adapted to the scoring environment. Find the necessary data for a sport of your choice and compute the optimal exponent to the Pythagorean formula. (The website www.opensourcesports.com provides databases for NBA and WNBA basketball and for NHL hockey in a format similar to Lahman's baseball database.)

5

Value of Plays Using Run Expectancy

CONTENTS

5.1	The Run Expectancy Matrix	105
5.2	Runs Scored in the Remainder of the Inning	106
5.3	Creating the Matrix	107
5.4	Measuring Success of a Batting Play	110
5.5	Albert Pujols	111
5.6	Opportunity and Success for All Hitters	114
5.7	Position in the Batting Lineup	116
5.8	Run Values of Different Base Hits	119
	5.8.1 Value of a home run	119
	5.8.2 Value of a single	121
5.9	Value of Base Stealing	123
5.10	Further Reading and Software	126
5.11	Exercises	126

5.1 The Run Expectancy Matrix

An important matrix in sabermetrics research is the run expectancy matrix. As each base (first, second, and third) can be occupied by a runner or empty, there are $2 \times 2 \times 2 = 8$ possible arrangements of runners on the three bases. The number of outs can be 0, 1, or 2 (three possibilities), and so there are a total of $8 \times 3 = 24$ possible arrangements of runners and outs. For each combination of runners on base and outs, one is interested in computing the average number of runs scored in the remainder of the inning. When these average runs are arranged as a table classified by runners and outs, the display is often called the run expectancy matrix. We illustrate using R to compute this matrix using play-by-play data for the 2011 season. This matrix is used to define the average run value (or run value) of a batter's plate appearance. Then the distribution of average run values is explored for all batters in the 2011 season. The run values for Albert Pujols are used to help understand the pattern of run values. We continue by exploring how players in different positions in the batting lineup perform with respect to this criterion. The notion of run value is helpful for understanding the relative benefit of different batting plays and we explore the value of a home run and a single. This chapter is concluded by using the run expectancy matrix and run values to understand the benefit of stealing a base and the cost of being caught stealing.

5.2 Runs Scored in the Remainder of the Inning

We begin by reading into R the play-by-play database for the 2011 season and storing this data into the R data frame `data2011`. The `fields.csv` file contains the names of the variables of this database and the `names` function is used to attach these variable names to the data frame.

```
data2011 <- read.csv("all2011.csv", header=FALSE)
fields <- read.csv("fields.csv")
names(data2011) <- fields[, "Header"]
```

At a given plate appearance, there is a potential to score runs. Clearly, this potential is greater with runners on base, specifically runners in scoring position (second or third base), and when there are few outs. This runs potential is measured by computing the average number of runs scored in the remainder of the inning for each combination of runners on base and number of outs. Certainly, the average runs scored is dependent on many variables such as home versus away, the current score, the pitching and the defense. But this runs potential represents the opportunity to create runs in a typical situation during an inning and is useful for measuring contributions of players in an average scenario.

To compute the number of runs scored in the remainder of the inning, we need to know the total runs scored by both teams at the plate appearance and also the total runs scored by the teams at the end of the specific half-inning. The runs scored in the remainder of the inning, denoted by RUNS.ROI, is the difference

$$RUNS.ROI = \text{Total Runs Scored in Inning} - \text{Current Runs Scored.}$$

To begin, a variable `RUNS` is created that is equal to the sum of the visitor's score (`AWAY_SCORE_CT`) and the home team's score (`HOME_SCORE_CT`) at each plate appearance.

```
data2011$RUNS <- with(data2011, AWAY_SCORE_CT + HOME_SCORE_CT)
```

A new variable `HALF.INNING` is also created, using the `paste` function, combining the game id, the inning, and the team at bat.

```
data2011$HALF.INNING <- with(data2011,
    paste(GAME_ID, INN_CT, BAT_HOME_ID))
```

The variable `HALF.INNING` creates a unique identification for each half-inning of every game during the season.

We wish to compute the maximum total score for each half-inning, combining home and visitor scores. To accomplish this, a new variable `RUNS.SCORED` is created that gives the number of runs scored for each play. (The variables `BAT_DEST_ID`, `RUN1_DEST_ID`, `RUN2_DEST_ID`, and `RUN3_DEST_ID` give the destination bases for the batter and each runner, and runs are scored for each destination base that exceeds 3.) By use of the `aggregate` function with the `sum` function, we compute the total runs scored in each half-inning and store in the data frame `RUNS.SCORED.INNING`.

```
data2011$RUNS.SCORED <- with(data2011, (BAT_DEST_ID > 3) +
  (RUN1_DEST_ID > 3) + (RUN2_DEST_ID > 3) + (RUN3_DEST_ID > 3))
RUNS.SCORED.INNING <- aggregate(data2011$RUNS.SCORED,
              list(HALF.INNING=data2011$HALF.INNING), sum)
```

By another application of the `aggregate` function with the "`[`" function, we find the total game runs at the beginning of each half-inning.

```
RUNS.SCORED.START <- aggregate(data2011$RUNS,
          list(HALF.INNING=data2011$HALF.INNING), "[", 1)
```

The maximum total score in a half-inning is the sum of the initial total runs and the runs scored. A new data frame `MAX` is created, defining a new variable x equal to the maximum runs scored. The `merge` function is used to merge this information with the data frame `data2011` and create the new maximum total score variable `MAX.RUNS`. (The function `ncol` gives the number of columns of a data frame.)

```
MAX <- data.frame(HALF.INNING=RUNS.SCORED.START$HALF.INNING)
MAX$x <- RUNS.SCORED.INNING$x + RUNS.SCORED.START$x
data2011 <- merge(data2011, MAX)
N <- ncol(data2011)
names(data2011)[N] <- "MAX.RUNS"
```

Now the runs scored in the remainder of the inning (new variable `RUNS.ROI`) can be computed by taking the difference of `MAX.RUNS` and `RUNS`.

```
data2011$RUNS.ROI <- with(data2011, MAX.RUNS - RUNS)
```

5.3 Creating the Matrix

Now that the runs scored in the remainder of the inning variable have been computed for each plate appearance, it is straightforward to compute the run expectancy matrix.

Currently, there are three variables BASE1_RUN_ID, BASE2_RUN_ID, and BASE2_RUN_ID containing the player codes of the baserunners (if any) who are respectively on first, second, or third base. Three new binary variables RUNNER1, RUNNER2, and RUNNER3 are created that are either 1 or 0 if the corresponding base is respectively occupied or empty. (The `as.character` function converts a factor variable to a character variable.)

```
RUNNER1 <- ifelse(as.character(data2011[ ,"BASE1_RUN_ID"]) == "", 0, 1)
RUNNER2 <- ifelse(as.character(data2011[ ,"BASE2_RUN_ID"]) == "", 0, 1)
RUNNER3 <- ifelse(as.character(data2011[ ,"BASE3_RUN_ID"]) == "", 0, 1)
```

A short function `get.state` is written to create a state variable, combining (using the function `paste`) the runner indicators and the number of outs. This function is used to create a current state variable STATE.

```
get.state <- function(runner1, runner2, runner3, outs){
  runners <- paste(runner1, runner2, runner3, sep="")
  paste(runners, outs)
}
data2011$STATE <- get.state(RUNNER1, RUNNER2, RUNNER3, data2011$OUTS_CT)
```

One particular state value would be "011 2" which indicates that there are currently runners on second and third base with two outs. A second state value "100 0" indicates there is a runner at first with no outs.

We want to only consider plays in our data frame where there is a change in the runners on base, number of outs, or the runs scored. Three new variables NRUNNER1, NRUNNER2, NRUNNER3 are created which indicate, respectively, if first base, second base, and third base are occupied after the play. (The function `as.numeric` converts a logical variable to a numeric variable.) The variable NOUTS is the number of outs after the play, and RUNS.SCORED is the number of runs scored on the play. Again the `get.state` function is used to create the variable NEW.STATE giving the runners on each base and the number of outs after the play.

```
NRUNNER1 <- with(data2011, as.numeric(RUN1_DEST_ID == 1 |
  BAT_DEST_ID == 1))
NRUNNER2 <- with(data2011, as.numeric(RUN1_DEST_ID == 2 |
  RUN2_DEST_ID == 2 | BAT_DEST_ID==2))
NRUNNER3 <- with(data2011, as.numeric(RUN1_DEST_ID == 3 |
  RUN2_DEST_ID == 3 | RUN3_DEST_ID == 3 | BAT_DEST_ID == 3))
NOUTS <- with(data2011, OUTS_CT + EVENT_OUTS_CT)
data2011$NEW.STATE <- get.state(NRUNNER1, NRUNNER2, NRUNNER3, NOUTS)
```

By use of the `subset` function, attention is restricted to plays where either there is a change between STATE and NEW.STATE (indicated by the not equal logical operator "!=") or there are runs scored on the play.

```
data2011 <- subset(data2011, (STATE != NEW.STATE) | (RUNS.SCORED > 0))
```

Before the runs expectancies are computed, one final adjustment is necessary. The play-by-play database includes scoring information for all half-innings during the 2011 season, including partial half-innings at the end of the game where the winning run is scored with less than three outs. In our work, we want to work only with complete half-innings where three outs are recorded. The `ddply` function in the `plyr` package is applied to compute the number of outs for each half-inning, and the `merge` function is used to add a new variable `Outs.Inning` to the data frame. The `subset` function is used to extract the data from the half-innings in `data2011` with exactly three outs – the new data frame is named `data2011C`. (By removing the noncomplete innings, one is introducing a small bias since these innings are not complete due to the scoring of at least one run.)

```
library(plyr)
data.outs <- ddply(data2011, .(HALF.INNING), summarize,
     Outs.Inning=sum(EVENT_OUTS_CT))
data2011 <- merge(data2011, data.outs)
data2011C <- subset(data2011, Outs.Inning == 3)
```

The expected number of runs scored in the remainder of the inning (the run expectancy) is computed for each of the 24 bases/outs situations by use of the `aggregate` function, grouping by `STATE` with the `mean` function.

```
RUNS <- with(data2011C, aggregate(RUNS.ROI, list(STATE), mean))
```

To display these run values as an 8×3 matrix, a new variable `Outs` is defined, and the `RUNS` data frame is sorted by the number of outs.

```
RUNS$Outs <- substr(RUNS$Group, 5, 5)
RUNS <- RUNS[order(RUNS$Outs), ]
```

The `matrix` function is used to create a matrix `RUNS.out`. For readability, each value is rounded to two decimal places and labels are assigned to the rows and columns by two applications of the `dimnames` function.

```
RUNS.out <-matrix(round(RUNS$x, 2), 8, 3)
dimnames(RUNS.out)[[2]] <- c("0 outs", "1 out", "2 outs")
dimnames(RUNS.out)[[1]] <- c("000", "001", "010", "011", "100", "101",
          "110", "111")
```

To see how the run expectancy values have changed over time, the 2002 season values as reported in Albert and Bennett (2003) are collected in the vector `RUNS.2002`. The 2011 and 2002 expectancies are displayed side-by-side for comparison purposes.

```
RUNS.2002 <- matrix(c(.51, 1.40, 1.14,  1.96, .90, 1.84, 1.51, 2.33,
          .27,  .94,  .68,  1.36, .54, 1.18,  .94, 1.51,
          .10,  .36,  .32,   .63, .23, .52,   .45, .78), 8, 3)
dimnames(RUNS.2002) <- dimnames(RUNS.out)
```

```
cbind(RUNS.out, RUNS.2002)
    0 outs 1 out 2 outs 0 outs 1 out 2 outs
000   0.47  0.25   0.10   0.51  0.27   0.10
001   1.45  0.94   0.32   1.40  0.94   0.36
010   1.06  0.65   0.31   1.14  0.68   0.32
011   1.93  1.34   0.54   1.96  1.36   0.63
100   0.84  0.50   0.22   0.90  0.54   0.23
101   1.75  1.15   0.49   1.84  1.18   0.52
110   1.41  0.87   0.42   1.51  0.94   0.45
111   2.17  1.47   0.76   2.33  1.51   0.78
```

It is somewhat remarkable that these run expectancy values have not changed over the recent history of baseball. That indicates that there have been little changes in the average run scoring tendencies of MLB teams between 2002 and 2011.

5.4   Measuring Success of a Batting Play

When a player comes to bat with a particular runners and outs situation, the run expectancy matrix tells us the average number of runs a team will score in the remainder of the half-inning. Based on the outcome of the plate appearance, the state (runners on base and outs) will change and there will be a updated run expectancy value. The value of the plate appearance, called the *run value*, is measured by the difference in run expectancies of the old and new states plus the number of runs scored on the particular play.

$$\text{RUNS VALUE} = \text{RUNS}_{New\ State} - \text{RUNS}_{Old\ State} + \text{RUNS}_{Scored\ on\ Play}$$

The run values for all plays in the original data frame `data2011` are computed using the following R script. First a 32×1 matrix `RUNS.POTENTIAL` is defined that contains the run expectancies for the 32 situations including 3 outs. The run expectancy of a situation with 3 outs is obviously zero, so we append some zeros to the previously calculated run expectancy values. The new variable `RUNS.STATE` is defined to be the run expectancy of the current state and the variable `RUNS.NEW.STATE` is defined to be the run expectancy of the new state. The new variable `RUNS.VALUE` is set equal to the difference in `RUNS.NEW.STATE` and `RUNS.STATE` plus the `RUNS.SCORED` variable.

```
RUNS.POTENTIAL <- matrix(c(RUNS$x, rep(0, 8)), 32, 1)
dimnames(RUNS.POTENTIAL)[[1]] <- c(RUNS$Group, "000 3", "001 3",
               "010 3", "011 3", "100 3", "101 3", "110 3", "111 3")
data2011$RUNS.STATE <- RUNS.POTENTIAL[data2011$STATE, ]
data2011$RUNS.NEW.STATE <- RUNS.POTENTIAL[data2011$NEW.STATE, ]
data2011$RUNS.VALUE <- data2011$RUNS.NEW.STATE - data2011$RUNS.STATE +
   data2011$RUNS.SCORED
```

5.5 Albert Pujols

To better understand run values, let's focus on the plate appearances for the great hitter Albert Pujols for the 2011 season. To find Pujols' player id, we read in the `roster2011.csv` data file and use the `subset` function to extract the `Player.ID` variable. The `as.character` function converts `Player.ID` from a factor to a character variable called `albert.id`.

```
Roster <- read.csv("roster2011.csv")
albert.id <- subset(Roster, First.Name == "Albert" &
                    Last.Name == "Pujols")$Player.ID
albert.id <- as.character(albert.id)
```

Using the `subset` function, a data frame of Pujols plate appearances is found, where the batter id (variable `BAT_ID`) is equal to `albert.id`.

```
albert <- subset(data2011, BAT_ID == albert.id)
```

We wish to consider only the batting plays where Albert was the hitter, so the `subset` function is used to select the rows where the batting flag (variable `BAT_EVENT_FL`) is true.[1]

```
albert <- subset(albert, BAT_EVENT_FL == TRUE)
```

How did Albert do on his first two plate appearances this season? To answer this, we display the first two rows of the data frame `albert`, showing the original state, new state, and run value variables:

```
albert[1:2, c("STATE", "NEW.STATE", "RUNS.VALUE")]
     STATE NEW.STATE RUNS.VALUE
6556 100 1   000 3 -0.4960492
6574 001 2   001 3 -0.3173913
```

On his first plate appearance, there was a runner on first with one out. The outcome of this plate appearance was three outs, indicating that Albert hit into a double-play, and the run value for this play was −0.496. On his second plate appearance, there was a runner on third with two outs. Evidently Albert got out (the final state had three outs) and the run value was −0.317. Based on the run values of these first plate appearances, Albert didn't have a very good start to the 2011 season.

When one evaluates the run values for any player, there are two primary questions. First, we need to understand the player's opportunities for producing runs. What were the runner/outs situations for the player's plate appearances? Second, what did the batter do with these opportunities to score runs?

---

[1]The variable `BAT_EVENT_FL` distinguishes batting events from non-batting events such as steals and wild pitches.

The batter's success or lack of success on these opportunities can be measured by use of the run values.

Let's focus on the runners states to understand Albert's opportunities. Since a few of the counts of the runners/outs states over the 32 outcomes are close to zero, we focus on runners on base and use the `substr` function to create a new variable `RUNNERS` containing the runners state of the three bases. The `table` function is applied to tabulate the runners state for all of Albert's plate appearances.

```
albert$RUNNERS <- substr(albert$STATE, 1, 3)
table(albert$RUNNERS)
```

```
000 001 010 011 100 101 110 111
354  27  60   7 135  19  36  13
```

We see that Albert generally was batting with the bases empty (`000`) or with only a runner on first (`100`). Most of the time, Albert was batting with no runners in scoring position.

How did Albert perform with these opportunities? Using the following R code, we construct a stripchart (using the `stripchart` function) that shows the run values for all plate appearances organized by the runners state. (See Figure 5.1.) A horizontal line at the value zero is added to the graph – points above the line (below the line) correspond to positive (negative) contributions.

```
with(albert, stripchart(RUNS.VALUE ~ RUNNERS, vertical=TRUE, jitter=0.2,
       xlab="RUNNERS", method="jitter", pch=1, cex=0.8))
abline(h=0)
```

There are many duplicate run values, so we jitter the points (using the `method="jitter"` argument) to better show the density of run values. When the bases were empty (`000`), the range of possible run values was relatively small. For this state, the large cluster of points at a negative run value corresponds to the many occurrences when Albert got an out with the bases empty. The cluster of points at (`000`) at the value 1 corresponds to Albert's home runs with the bases empty. (A home run with runners empty will not change the bases/outs state and the value of this play is exactly one run.) For other situations, say the bases-loaded situation (`111`), there is much variation in the run values. For one plate appearance, the state moved from `111 1` to `000 3`, indicating that Albert hit into a inning-ending double play with the bases loaded with a run value of -1.41. In contrast, Albert did hit a home run with the bases loaded with no outs and the run value of this outcome was 2.38. (The run value of a grand slam is not 4 since the run potential of the end state of bases empty is much smaller than the run potential of a bases-loaded state.)

To understand Albert's total run production for the 2011 season, the `aggregate` function together with the `sum` and `length` functions can be used to compute the number of opportunities and sum of run values for each of the runners situations.

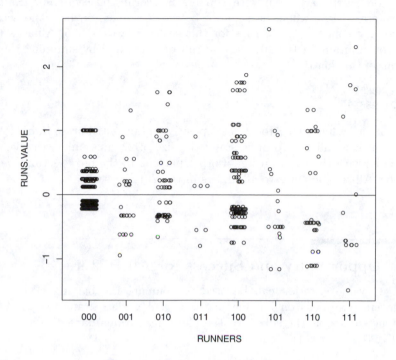

FIGURE 5.1
Stripchart of run values of Albert Pujols for all 2011 plate appearances as a function of the runners state. The points have been jittered since there are many plate appearances with identical run values.

```
A.runs <- aggregate(albert$RUNS.VALUE, list(albert$RUNNERS), sum)
names(A.runs)[2] <- "RUNS"
A.PA <- aggregate(albert$RUNS.VALUE, list(albert$RUNNERS), length)
names(A.PA)[2] <- "PA"
A <- merge(A.PA, A.runs)
A
   Group.1  PA        RUNS
1      000 354  10.7568515
2      001  27  -0.9417766
3      010  60   9.6399002
4      011   7  -0.5680439
5      100 135  14.2174870
6      101  19  -1.9853612
7      110  36  -4.5593651
8      111  13   0.6971003
```

We see, for example, that Albert came to bat with the runners empty 354 times, and his total run value contribution to these 354 PAs was 10.76. Albert didn't do particularly well with runners in scoring position. For example, there were 36 PAs where he came to bat with runners on first and second, and his net contribution in runs for this situation was −4.56. Albert's total runs contribution for the 2011 season can be computed by summing the last column of this matrix.

```
sum(A$RUNS)
[1] 27.25679
```

It is not surprising that Albert has a positive total contribution in his PAs in 2011, but it is difficult to understand the size of 27 runs unless this value is compared with the contribution of other players. In the next section, we will see how Albert compares to all hitters in the 2011 season.

5.6   Opportunity and Success for All Hitters

The run value criterion can be used to compare the batting effectiveness of players. We focus on batting plays, so a new data frame `data2011b` is constructed that is the subset of the main data frame `data2011` where the `BAT_EVENT_FL` variable is equal to `TRUE`:

```
data2011b <- subset(data2011, BAT_EVENT_FL == TRUE)
```

It is difficult to compare the total run values of two players on face value, since they have different opportunities to create runs for their teams. One player in the middle of the batting order may come to bat many times when there are runners in scoring position and good opportunities to create runs. Other players towards the bottom of the batting order may not get the same opportunities to bat with runners on base. One can measure a player's opportunity to create runs by the sum of the runs potential state (variable `RUNS.STATE`) over all of his plate appearances. We can summarize a player's batting performance in a season by the total number of plate appearances, the sum of the runs potentials, and the sum of the run values.

The R function `aggregate` is helpful in obtaining these summaries. In the following R code, the data frame `runs.pa` contains the number of plate appearances for all batters in the 2011 season, the data frame `runs.sums` contains the total run value for these players, and the data frame `runs.start` contains the total starting runs potential for the players. Using two applications of the `merge` function, we merge the two matrices, creating the data frame `runs`. A row of this data frame will contain the number of plate appearances, the total run value, and the total run potential for a particular player.

```
runs.sums <- aggregate(data2011b$RUNS.VALUE, list(data2011b$BAT_ID), sum)
runs.pa <- aggregate(data2011b$RUNS.VALUE, list(data2011b$BAT_ID), length)
runs.start <- aggregate(data2011b$RUNS.STATE, list(data2011b$BAT_ID), sum)
```

```
names(runs.sums) <- c("Batter", "Runs")
names(runs.pa) <- c("Batter", "PA")
names(runs.start) <- c("Batter", "Runs.Start")
runs <- merge(runs.sums, runs.pa)
runs <- merge(runs, runs.start)
```

The data frame `runs` contains batting data for both pitchers and nonpitchers. It seems reasonable to restrict attention to nonpitchers, since pitchers and nonpitchers have very different batting abilities. Also we limit our focus on the players who are primarily starters on their teams. One can remove pitchers and nonstarters by focusing on batters with at least 400 plate appearances. A new data frame `runs400` is created by an application of the `subset` function. There are 203 players in this data frame; we display the first few rows by use of the `head` function.

```
runs400 <- subset(runs, PA >= 400)
head(runs400)
      Batter      Runs  PA Runs.Start
1   abreb001  9.695694 585   251.6928
15  andir001 -8.511571 511   246.0340
16  andre001  4.847767 665   323.6004
18  ankir001 -4.326331 415   183.3090
19  arenj001  6.358087 486   226.9625
24  avila001 32.125231 551   266.5247
```

Is there a relationship between batters' opportunities and their success in converting these opportunities to runs? To answer this question, we construct a scatterplot of run opportunity (`Runs.Start`) against run value (`Runs`) for these hitters with at least 400 at bats (see Figure 5.2). To help see the pattern in this scatterplot, the `lowess` function is used to smooth this scatterplot and the smoothing curve is placed on top of the scatterplot by use of the `lines` function. To interpret this graph, it is helpful to add a horizontal line (using the `abline` function) at `Runs=0`; points above this line correspond to hitters who had a total positive run value contribution in the 2011 season.

```
with(runs400, plot(Runs.Start, Runs))
with(runs400, lines(lowess(Runs.Start, Runs)))
abline(h=0)
```

From viewing Figure 5.2, we see that batters with larger values of `Runs.Start` tend to have larger runs contributions. But there is a wide spread in the run values for these players. In the group of players who have `Runs.Start` values between 300 and 350, four of these players actually have negative runs contributions and other players created over 60 runs in the 2011 season.

From the graph, we see that only a limited number of players created more than 40 runs for their teams. Who are these players? In the R code, we create a new data frame `runs400.top` containing the runs statistics for only the players who created more than 40 runs. For labeling purposes, we would like to obtain the last names of the players available on the "roster2011.csv" data file. We read in this file using the `read.csv` function and use the `merge` function to merge the roster information with the data frame `runs400.top`. By use of the `text` function, point labels are added to the previous scatterplot for these outstanding hitters. (See Figure 5.2.)

```
runs400.top <- subset(runs400, Runs >= 40)
roster2011 <- read.csv("roster2011.csv")
runs400.top <- merge(runs400.top,
            roster2011, by.x="Batter", by.y="Player.ID")
with(runs400.top, text(Runs.Start, Runs, Last.Name, pos=1))
```

From this figure, we learn that the best hitters using the runs criterion are Miguel Cabrera (71.11), Jose Bautista (67.37), Prince Fielder (60.82), Joey Votto (60.65), and Matt Kemp (59.64). There is an interesting outlier in this figure – Mike Napoli created over 40 runs for his team despite only having a `Runs.Start` value close to 200. Napoli was a very productive batter for his team given his opportunities to produce runs.

5.7   Position in the Batting Lineup

Managers like to put their best hitters in the middle of the batting lineup. Traditionally, a team's "best hitter" bats third and the clean-up hitter in the fourth position is the best batter for advancing runners on base. What are the batting positions of the hitters in our sample? Specifically, are the best hitters using the run value criterion the ones who bat in the middle of the lineup?

A player may bat in several positions in the lineup during the season. We define a player's batting position as the position that he bats most frequently. A function `get.batting.pos` is defined that will find a player's batting position. This function will compute a table of a player's batting position and identify the position which has the highest frequency. By use of the `sapply` function together with the function `get.batting.pos`, we find the batting position for all players in the `runs400` data frame.

```
get.batting.pos <- function(batter){
  TB <- table(subset(data2011, BAT_ID == batter)$BAT_LINEUP_ID)
  names(TB)[TB == max(TB)][1]}
position <- sapply(as.character(runs400$Batter), get.batting.pos)
```

In the following R code, the players' run opportunities are plotted against their run values. By use of the `type="n"` option, the axes are drawn but points are not plotted. Instead, we use the `text` function with labels contained in the vector `position` to display the batting positions as plotting points. (See Figure 5.3.)

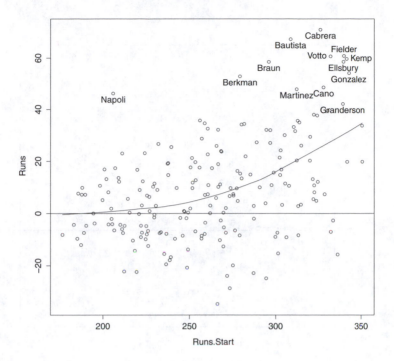

FIGURE 5.2
Scatterplot of total run value against the runs potential for all players in the 2011 season with at least 400 plate appearances. A smoothing curve is added to the scatterplot – this shows that players who had more run potential tend to have large run values. The players with a total run value at least 40 are labeled.

```
with(runs400, plot(Runs.Start, Runs, type="n"))
with(runs400, lines(lowess(Runs.Start, Runs)))
abline(h=0)
with(runs400, text(Runs.Start, Runs, position))
```

From this figure, we better understand the relationship between batting position, run opportunities, and run values. The best hitters, the ones who create a large number of runs, generally bat third, fourth, and fifth in the batting order. The number of runs created by the leadoff (first) and second batters in the lineup are much smaller than the runs created by the best hitters in the middle (third and fourth positions) of the lineup. There are some surprises from this general pattern of batting positions. Mike Napoli, the unusual hitter who created over 40 runs with only 200 run opportunities, bats only sixth in the lineup. Also, there are many cleanup hitters displayed who have mediocre values of runs created.

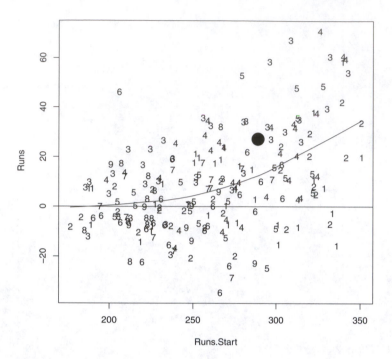

FIGURE 5.3
Scatterplot of total run value against the runs potential for all players in the
2011 season with at least 400 plate appearances. The points are labeled by
the position in the batting lineup and the large point corresponds to Albert
Pujols.

How does Albert Pujols and his total run value of 27.2 compare among the
group of hitters with at least 400 plate appearances? By use of the `subset`
function, we find Pujols' data from the `runs400` matrix. Using the `points`
function, we display Albert's (`Runs.Start`, `Runs`) value by a large solid dot.
In this particular season (2011), Albert was one of the better hitters in terms
of creating runs for his team.

```
AP <- subset(runs400, Batter == albert.id)
points(AP$Runs.Start, AP$Runs, pch=19, cex=3)
```

5.8 Run Values of Different Base Hits

There are many applications of run values in studying baseball. Here we look at the values of a home run and a single from the perspective of creating runs.

One criticism of a batting average is that it gives equal value to the four possible base hits (single, double, triple, and home run). One way of distinguishing the values of the base hits is to assign the number of bases reached – 1 for a single, 2 for a double, 3 for a triple, and 4 for a home run. The slugging percentage is the total number of bases divided by the number of at-bats. But it is not clear that the values 1, 2, 3, and 4 represent a reasonable measure of the value of the four possible base hits. We can get a better measure of the importance of these base hits by the use of run values.

5.8.1 Value of a home run

Let's focus on the value of a home run from a runs perspective. The home run plays are extracted from the data frame `runs2011` using the `EVENT_CD` play event variable. A value of `EVENT_CD` of 23 corresponds to a home run. Using the `subset` function with the `EVENT_CD == 23` condition, a new data frame `d.homerun` is created with the home run plays.

```
d.homerun <- subset(data2011, EVENT_CD == 23)
```

When are the runners/outs states for the home runs hit during the 2011 season? We answer this question by use of the `table` function.

```
table(d.homerun$STATE)
```

```
000 0 000 1 000 2 001 0 001 1 001 2 010 0 010 1 010 2 011 0 011 1
 1226   812   630    15    50    54    57    94   144    17    28
011 2 100 0 100 1 100 2 101 0 101 1 101 2 110 0 110 1 110 2 111 0
   32   262   301   286    32    50    61    60   115   128    15
111 1 111 2
   42    41
```

By use of the `prop.table` function, the relative frequencies are computed and the `round` function is used to round the values to three decimal spaces.

```
round(prop.table(table(d.homerun$STATE)), 3)
```

```
000 0 000 1 000 2 001 0 001 1 001 2 010 0 010 1 010 2 011 0 011 1
0.269 0.178 0.138 0.003 0.011 0.012 0.013 0.021 0.032 0.004 0.006
011 2 100 0 100 1 100 2 101 0 101 1 101 2 110 0 110 1 110 2 111 0
0.007 0.058 0.066 0.063 0.007 0.011 0.013 0.013 0.025 0.028 0.003
111 1 111 2
0.009 0.009
```

We see from this table that the fraction of home runs hit with the bases empty is $0.269 + 0.178 + 0.138 = 0.585$. So over half of the home runs are hit with no runners on base.

What are the run values of these home runs? We already observed in the analysis of Pujols' data that the run value of a home run with the bases empty is one. A histogram of the run values for all home runs is constructed using the truehist function in the MASS package.[2] (See Figure 5.4.)

```
library(MASS)
truehist(d.homerun$RUNS.VALUE)
```

It is obvious from this graph that most home runs (the ones with the bases empty) have a run value of one. But there is a cluster of home runs with values between 1.5 and 2.0, and there is a small group of home runs with run value exceeding three. Which runners/outs situation leads to the most

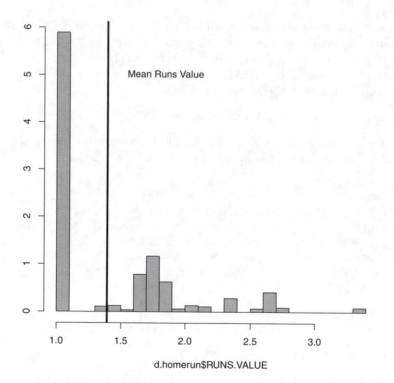

FIGURE 5.4
Histogram of the run values of the home runs hit during the 2011 season. The vertical line shows the location of the mean run value of a home run.

---

[2]We prefer truehist to the hist function since by default density values, instead of counts, are graphed and it has a nice appearance.

valuable home runs? Using the `subset` function, the row of the data frame is extracted corresponding to the largest run value.

```
subset(d.homerun, RUNS.VALUE == max(RUNS.VALUE))[1,
          c("STATE", "NEW.STATE", "RUNS.VALUE")]
      STATE NEW.STATE RUNS.VALUE
6748 111 2    000 2    3.336175
```

As one might expect, the most valuable home run occurs when there are bases loaded with two outs. From our earlier work, it was seen that this type of home run occurred 41 times during the season, and the run value of this home run is 3.34.

Overall, what is the run value of a home run? This question is answered by computing the run value of all the home runs in the data frame `d.homerun`.

```
mean.HR <- mean(d.homerun$RUNS.VALUE)
mean.HR
[1] 1.392393
```

A vertical line is drawn on the graph showing the mean run value and a label is added to this line. (See Figure 5.4.)

```
abline(v=mean.HR, lwd=3)
text(1.5, 5, "Mean Run Value", pos=4)
```

This average run value is pretty small, but this value partially reflects the fact that most home runs are hit with the bases empty.

5.8.2 Value of a single

Run values can also be used to evaluate the benefit of a single. Unlike a home run, the run value of a single will depend both on the initial state (runners and outs) and on the final state. The final state of a home run will always have the bases empty; in contrast, the final state of a single will depend on the movement of any runners on base.

In R, the `subset` function is used to select the plays where EVENT_CD == 20 (corresponding to a single); the new data frame is called `d.single`. A histogram is constructed of the run values for all of the singles in the 2011 season. (See Figure 5.5.)

```
d.single <- subset(data2011, EVENT_CD == 20)
library(MASS)
truehist(d.single$RUNS.VALUE)
```

Looking at the histogram of run values of the single, there are three large spikes between 0 and 0.5. These large spikes can be explained by constructing a frequency table of the beginning state.

```
table(d.single$STATE)
```

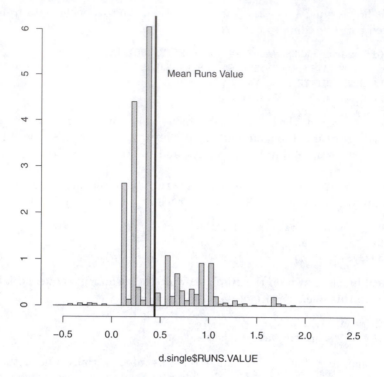

FIGURE 5.5
Histogram of the run values of the singles hit during the 2011 season. The vertical line shows the location of the mean run value of a single.

```
000 0 000 1 000 2 001 0 001 1 001 2 010 0 010 1 010 2 011 0 011 1
7078  5067  3814    85   315   356   502   844   959    87   225
011 2 100 0 100 1 100 2 101 0 101 1 101 2 110 0 110 1 110 2 111 0
 216  1593  2063  1826   154   344   385   346   728   770   100
111 1 111 2
 284   277
```

We see that most of the singles occur with the bases empty, and the three spikes in the histogram, as one moves from left to right in Figure 5.5, correspond to singles with no runners on and two outs, one out, and no outs. The small cluster of run values in the interval 0.5 to 2.0 correspond to singles hit with runners on base.

What is the most valuable single from the run value perspective? We use the `subset` function to find the beginning and end states for the single that resulted in the largest run value.

```
subset(d.single, d.single$RUNS.VALUE ==
    max(d.single$RUNS.VALUE))[ , c("STATE", "NEW.STATE", "RUNS.VALUE")]
```

```
        STATE NEW.STATE RUNS.VALUE
105749 111 2     001 2    2.556382
```

In this particular play, the hitter came to bat with the bases loaded and two outs, and the final state was a runner on third with two outs. How could have this happened with a single? The data frame does contain a brief description of the play. But from the data frame we identify the play happening during the bottom of the 7th inning of a game between the Brewers and Twins on July 3, 2011. We check with www.espn.com to find the following play description. "D Valencia singled to deep left, J Mauer and M Cuddyer scored, J Thome to second, J Thome scored, D Valencia to third on error by left fielder M Kotsay." So evidently, the left fielder made an error on the fielding of the single that allowed all three runners to score and the batter to reach third base.

At the other extreme, by use of the subset function, two plays are identified which achieved the smallest run value.

```
subset(d.single, d.single$RUNS.VALUE == min(d.single$RUNS.VALUE))[
   , c("STATE", "NEW.STATE", "RUNS.VALUE")]
        STATE NEW.STATE RUNS.VALUE
69618  010 0     100 1 -0.5622312
138351 010 0     100 1 -0.5622312
```

How could the run value of a single be negative one-half a run? With further investigation, we find that in each case, there was a runner on second who was thrown out at the plate as a result of the single.

As in the case of the home run, it is straightforward to compute the mean run value of a single. We display this mean value on the histogram in Figure 5.5.

```
mean.single <- mean(d.single$RUNS.VALUE)
mean.single
[1] 0.4424186
abline(v=mean.single, lwd=3)
text(.5, 5, "Mean Run Value", pos=4)
```

In this case, we see that the mean value of a single is approximately equal to the run value when a single is hit with the bases empty with no outs. It is interesting that the run value of a single can be large (in the 1 to 2 range). These large run values reflect the fact that the benefit of the single depends on the advancement of the runners.

5.9   Value of Base Stealing

The run expectancy matrix is also useful in understanding the benefits of stealing bases. When a runner attempts to steal a base, there are two likely outcomes – either the runner will be successful in stealing the base or the runner will be caught stealing. Overall, is there a net benefit to attempting to steal a base?

The variable EVENT_CD gives the code of the play and codes of 4 and 6 correspond respectively to a stolen base (SB) or caught stealing (CS). Using the subset function, a new data frame stealing is created that consists of only the plays where a stolen base is attempted.

```
stealing <- subset(data2011, EVENT_CD == 6 | EVENT_CD == 4)
```

By use of the table function, we find the frequencies of the SB and CS outcomes.

```
table(stealing$EVENT_CD)
```

```
   4    6
2863  864
```

Among all stolen base attempts, the proportion of stolen bases is 2863 /(2863 + 864) = 0.768.

What are common runners/outs situations for attempting a stolen base? This is answered by constructing a frequency table for the STATE variable.

```
table(stealing$STATE)
```

```
001 1 001 2 010 0 010 1 010 2 011 1 100 0 100 1 100 2 101 0 101 1
   10     1    19   114   101     1   758  1018  1160    35   111
101 2 110 0 110 1 110 2 111 1
  180    27   102    89     1
```

We see that stolen base attempts typically happen with a runner only on first (state "100"). But there is a wide variety of situations where runners attempt to steal.

Every stolen base attempt has a corresponding run value that is stored in the variable RUNS.VALUE. This run value reflects the success of the attempt (either SB or CS) and the situation (runners and outs) where this attempt occurs. Using the truehist function, a histogram is constructed of all of the runs created for all the stolen base attempts.

```
library(MASS)
truehist(stealing$RUNS.VALUE)
```

Generally, all of the successful SBs have positive run value, although most of the values fall in the interval from 0 to 0.3. In contrast, the unsuccessful CSs (as expected) have negative run values. In further exploration, one can show the three spikes for negative run values correspond to CS when there is only a runner on first with 0, 1, and 2 outs.

Let's focus on the benefits of stolen base attempts in a particular situation. We create a new data frame which gives the attempted stealing data when there is a runner on first base with one out (state "100 1").

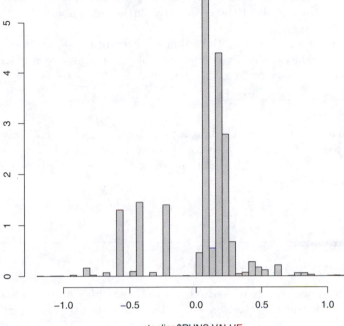

FIGURE 5.6
Histogram of the run values of all steal attempts during the 2011 season.

```
stealing.1001 <- subset(stealing, STATE == "100 1")
```

By tabulating the EVENT_CD variable, we see the runner successfully stole 753 times out of 753 + 265 attempts for a success rate of 74.0%.

```
table(stealing.1001$EVENT_CD)

  4   6
753 265
```

Another way to look at the outcome is to look at the frequencies of the NEW_STATE variable.

```
with(stealing.1001, table(NEW.STATE))
NEW.STATE
000 2 001 1 010 1
  262    52    704
```

This provides more information than simply recording a stolen base. On 704 occurrences, the runner successfully advanced to second base. On an additional 52 occurrences, the runner advanced to third. Perhaps this extra base was due to a bad throw from the catcher or a misplay by the infielder; more can be learned about the details of these plays by further examination of the other variables.

We are most interested in the value of attempting stolen bases in this situation – we address this by computing the mean run values of all of the attempts with a runner on first with one out.

```
mean(stealing.1001$RUNS.VALUE)
[1] 0.02649315
```

Stolen base attempts are worthwhile, although the value overall is about 0.02 runs per attempt. Of course, the actual benefit of the attempt depends on the success or failure and on the situation (runners and outs) where the stolen base is attempted.

5.10   Further Reading and Software

Lindsey (1963) was the first researcher to analyze play-by-play data in the manner described in this chapter. Using data collected by his father for the 1959-60 season, Lindsey obtained the run expectancy table that gives the average number of runs in the remainder of the inning for each of the runners/outs situations. Chapters 7 and 9 of Albert and Bennett (2003) illustrate the use of the run expectancy table to measure the value of different base hits and to assess the benefits of stealing and sacrifice hits. Dolphin et al. (2007), in their Toolshed chapter, describe the run expectancy table as one of the fundamental tools used throughout their book. Also, run expectancy plays a major role in the essays in Keri et al. (2007).

Ben Baumer and Gregory Matthews have developed an R package `openWAR` providing a convenient way of reading play-by-play data from the web and computing run expectancies. This package uses these run expectancies in the computation of WAR (wins above replacement) measures for players. The website `www.fangraphs.com/library/misc/war/` introduces WAR, a useful way of summarizing a player's total contribution to his team.

5.11   Exercises

1. (**Run Values of Hits**)

    In Section 5.8, we found the average run value of a home run and a single.

    (a) Use similar R code as described in Section 5.8 for the 2011 season data to find the mean run values for a double, and for a triple.

(b) Albert and Bennett (2001) use a regression approach to obtain the weights 0.46, 0.80, 1.02, and 1.40 for a single, double, triple, and home run, respectively. Compare the results from Section 5.8 and part (a) with the weights of Albert and Bennett.

2. (**Value of Different Ways of Reaching First Base**)

There are three different ways for a runner to get on base, a single, walk (BB), or hit-by-pitch (HBP). But these three outcomes have different run values due to the different advancement of the runners on base. Use run values based on data from the 2011 season to compare the benefit of a walk, a hit-by-pitch, and a single when there is a single runner on first base.

3. (**Comparing Two Players with Similar OBPs.**)

Rickie Weeks (batter id "weekr001") and Michael Bourne (batter id "bourm001") both were leadoff hitters during the 2011 season. They had similar on-base percentages – .350 for Weeks and .349 for Bourne. By exploring the run values of these two payers, investigate which player was really more valuable to his team. Can you explain the difference in run values in terms of traditional batting statistics such as AVG, SLG, or OBP?

4. (**Create Probability of Scoring a Run Matrix**)

In Section 5.3, the construction of the run expectancy matrix from 2011 season data was illustrated. Suppose instead that one was interested in computing the proportion of times when at least one run was scored for each of the 24 possible bases/outs situations. Use R to construct this probability of scoring matrix.

5. (**Runner Advancement with a Single**)

Suppose one is interested in studying how runners move with a single.

(a) Using the `subset` function, select the plays when a single was hit. (The value of `EVENT_CD` for a single is 20.) Call the new data frame `d.single`.

(b) Use the `table` function with the data frame `d.single` to construct a table of frequencies of the variables `STATE` (the beginning runners/outs state) and `NEW.STATE` (the final runners/outs state).

(c) Suppose there is a single runner on first base. Using the table from part (b), explore where runners move with a single. Is it more likely for the lead runner to move to second, or to third base?

(d) Suppose instead there are runners on first and second. Explore where runners move with a single. Estimate the probability a run is scored on the play.

6. (**Hitting Evaluation of Players by run values**)

Choose several players who were good hitters in the 2011 season. For each player, find the run values and the runners on base for all plate appearances. As in Figure 5.1, construct a graph of the run values against the runners on base. Was this particular batter successful when there were runners in scoring position?

6

Advanced Graphics

CONTENTS

6.1	Introduction	129
6.2	The `lattice` Package	130
	6.2.1 Introduction	130
	6.2.2 The `verlander` dataset	130
	6.2.3 Basic plotting with `lattice`	132
	6.2.4 Multipanel conditioning	133
	6.2.5 Superposing group elements	134
	6.2.6 Scatterplots and dot plots	135
	6.2.7 The `panel` function	137
	6.2.8 Building a graph, step-by-step	139
6.3	The `ggplot2` Package	144
	6.3.1 Introduction	144
	6.3.2 The `cabrera` dataset	145
	6.3.3 The first layer	146
	6.3.4 Grouping factors	148
	6.3.5 Multipanel conditioning (faceting)	149
	6.3.6 Adding elements	150
	6.3.7 Combining information	151
	6.3.8 Adding a smooth line with error bands	151
	6.3.9 Dealing with cluttered charts	153
	6.3.10 Adding a background image	155
6.4	Further Reading	157
6.5	Exercises	157

6.1 Introduction

Chapter 3 introduced graphics in R, illustrating a variety of displays with functions provided by the `graphics` package. While this traditional package has sufficient flexibility for many purposes, the functions are more difficult to use in graphing complicated data structures such as associations among three variables. In this chapter, the `lattice` and the `ggplot2` packages are introduced that are well-suited for constructing more sophisticated graphical

129

displays. In this chapter we focus on the use of these packages for baseball data.

The datasets used in this chapter are contained in an R workspace, a file with extension `.Rdata` which can be loaded with the following command line:

```
load("data/balls_strikes_count.Rdata")
```

6.2   The `lattice` Package

6.2.1   Introduction

The `lattice` graphics system, written by Deepayan Sarkar, provides an R implementation of *Trellis* displays, a regular gridlike framework for visualizing multivariate data, popularized by William Cleveland. The `lattice` package is currently included with the R base distribution and it becomes available for use by loading with the `library` function.

```
library(lattice)
```

This package produces color figures by default, so the output one obtains by typing the code in this chapter differs from the black and white figures printed in these pages. (One who wishes to exactly reproduce the black and white graphs should change the `lattice` default theme with the following command: `trellis.par.set(canonical.theme(color=FALSE))`.)

6.2.2   The `verlander` dataset

The `lattice` package is illustrated using data from the `verlander` data frame containing five years of PITCHf/x data for 2011 Cy Young Award and MVP recipient Justin Verlander. To give a glimpse of the contents of `verlander`, a random sample of 20 rows is selected from the data frame in the following R code. The function `sample` selects a random sample of size 20 from a set of integers 1 to `nrow(verlander)`, where the function `nrow` gives the number of rows of a data frame. The bracket notation is used to choose the rows of the data frame `verlander` corresponding to the random integer values in `sampleRows`.[1]

```
sampleRows <- sample(1:nrow(verlander), 20)
verlander[sampleRows,]
```

The sample rows are reported in tabular form in Table 6.1.

---

[1] Further arguments in `sample` that can be specified are `replace`, which indicates whether sampling should be performed with replacement and is set as `FALSE` by default, and a vector `probs` if one wants to give different selection probabilities for the set of integers.

TABLE 6.1

A random sample of twenty rows of the verlander dataset.

(row#)	season	gamedate	pitch_type	balls	strikes	pitches	speed	px	pz	batter_hand
7924	2011	2011-04-11	FF	1	0	68	95.80	0.35	3.32	L
11850	2012	2012-09-14	CU	1	2	39	81.20	-1.05	1.11	L
3028	2009	2009-08-30	FT	1	2	94	95.90	0.14	2.43	L
10363	2011	2011-07-31	FF	1	1	64	98.50	-0.45	2.99	R
2890	2009	2009-08-24	FF	2	1	65	99.00	0.10	2.85	R
4370	2010	2010-04-27	FT	0	2	58	98.40	-0.05	4.47	L
8117	2011	2011-04-22	CH	0	1	27	89.00	0.77	1.21	L
11774	2012	2012-09-29	CH	1	1	2	85.20	-1.18	2.04	L
13308	2012	2012-07-04	CU	0	2	86	79.70	-1.63	1.79	L
4147	2010	2010-04-17	FT	0	2	61	97.30	-0.53	4.16	R
12223	2012	2012-09-08	SL	2	0	8	85.60	1.43	1.65	L
10609	2011	2011-08-11	CU	1	1	86	79.30	-0.44	2.99	R
9500	2011	2011-06-25	FF	0	2	49	97.00	0.65	3.88	R
326	2009	2009-04-22	FF	0	0	42	93.80	-0.42	3.63	R
10065	2011	2011-07-21	FF	1	1	17	97.70	0.67	4.11	L
8611	2011	2011-05-13	CU	0	0	80	78.70	-1.09	2.44	R
1795	2009	2009-07-01	FT	0	0	96	97.10	-0.21	2.28	L
13397	2012	2012-06-29	SL	1	1	116	84.80	0.52	3.75	R
7228	2010	2010-09-12	SL	1	0	82	87.00	0.61	2.24	R
882	2009	2009-05-20	CU	2	2	30	82.60	-0.48	1.51	L

The first two columns of the `verlander` data frame, `season` and `gamedate`, contain the season and the date in which the game was played. The `pitch_type` column indicates the type of pitch thrown by the pitcher as a two-character abbreviation. The five pitches in Verlander's repertoire are the four-seam fastball (`FF`), the two-seam fastball (`FT`), the curveball (`CU`), the change-up (`CH`), and the slider (`SL`.) The `balls` and `strikes` columns report the ball-strike count on the batter when the pitch was delivered and the `pitches` column contains the number of pitches already thrown by the pitcher in that particular game. Reading the first row in Table 6.1, we see that Verlander threw a four-seam fastball, delivered as the 69th pitch in the game, with a 1-0 count on the batter. The next three columns include PITCHf/x data: `speed` is the recorded speed at the release of the pitch, `px` and `pz` define the location of the pitch when it crosses the front of the plate. For example, the first pitch in the table crossed the plate 0.35 feet to the right of the middle of the plate, 3.32 feet from the ground. The final column indicates whether the opposing player was batting from the left or the right side of the plate.

6.2.3   Basic plotting with `lattice`

The `lattice` package can be used to construct the basic statistical graphs described in Chapter 3. For example, suppose we are interested in exploring the speeds of Verlander's pitches. The `histogram` function in the `lattice` package constructs a histogram analogous to the `hist` function in the base R package. For functions in the `lattice` package, the first argument is a formula usually requiring at least one variable on the left side of the tilde character (to be plotted on the y-axis) and one variable on the right side of the tilde (to be plotted on the x-axis). The `data` argument gives the data frame containing the variables. For a one-variable graph such as a histogram only one variable (such as `speed`) is needed in the formula.

```
histogram(~ speed, data=verlander)
```

The graphical display is shown in Figure 6.1(a).

The `densityplot` function constructs a density plot (Figure 6.1(b)), an alternative method of displaying the distribution of a single variable. The code to produce a density plot is similar to that of the histogram; the option `plot.points=FALSE` prevents the display of the data points at the base of the curve.

```
densityplot(~ speed, data=verlander, plot.points=FALSE)
```

The histogram and density plot give similar messages about the speeds of Verlander's pitches. There are two noticeable humps in the distribution; the smaller hump in the mid-80s corresponds to the speeds of the off-speed pitches and the larger hump in the mid-90s corresponds to the fastballs.

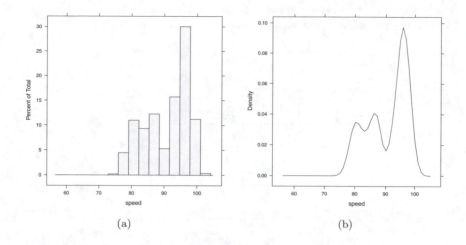

(a)   (b)

FIGURE 6.1
Histogram and density plot of Verlander's pitch speeds using functions in
`lattice` package.

6.2.4   Multipanel conditioning

One of the main enhancements of the `lattice` package over the traditional
`graphics` package is that it allows *multipanel conditioning*. In other words,
given a grouping factor, it is straightforward to produce a set of as many
displays corresponding to the groups identified by the factor.

As an example, suppose one is interested in constructing separate density
plots for the speed of Verlander's pitches where each plot is based on the speeds
of a particular pitch type. In this case, the grouping factor is `pitch_type` and,
by use of the `lattice` package, one can display these "conditional" density
plots on a series of panels using the same scales for the horizontal and vertical
variables.

The multipanel conditioning is obtained in the R graphing function by
adding to the formula the pipeline (|) symbol, followed by the conditioning
variable. For example, the set of conditional density plots is constructed by
means of the following command.[2]

```
densityplot(~ speed | pitch_type, data=verlander,
    layout=c(1, 5), plot.points=FALSE)
```

The argument option `plot.points=FALSE` suppresses the plotting of the data
on the horizontal axes. In Figure 6.2 density plots are plotted for each of the
five different pitch types composing Justin Verlander's arsenal. This figure

---

[2]The `layout` argument is used to indicate how the panels are to be arranged. A two-
element vector is specified, indicating the number of columns and rows, respectively.

clearly shows that Verlander's fastballs are thrown in the mid-90s, his sliders and changeups in the mid-80s, and his curveballs about 80 mph.

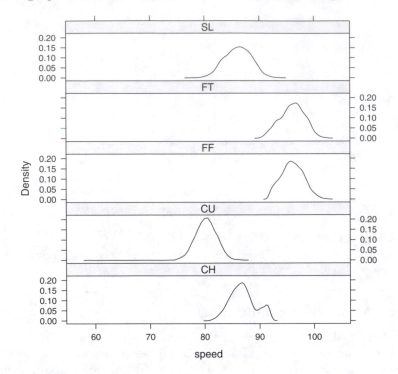

FIGURE 6.2
Panels of density plots of Verlander's pitch speed by pitch type.

6.2.5 Superposing group elements

An alternative way to graphically compare the pitch speeds by pitch type is to plot the lines of the density plots on a single panel, where one uses different line types or colors to distinguish the groups. To obtain this superposed display, one uses the `densityplot` function with the `groups` argument and the grouping variable `pitch_type`. The resulting display is shown in Figure 6.3.

```
densityplot(~ speed, data=verlander, groups=pitch_type,
    plot.points=FALSE, auto.key=TRUE)
```

For a reader to identify the pitch type for each line of the plot, a legend is needed. The `auto.key=TRUE` line automatically creates a legend on top of the plot region. This superposed display gives a similar message as the multipanel display – Verlander throws pitches at three different speeds.

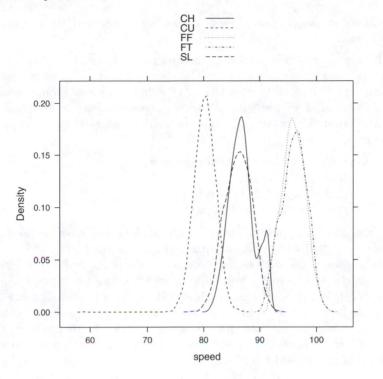

FIGURE 6.3
Density plot of Verlander's pitch speed by pitch type using superposed lines.

6.2.6 Scatterplots and dot plots

Multipanel conditioning plots and superposed plots can be constructed using a variety of graphical methods in the `lattice` package. The application of the `histogram` and `densityplot` functions have already been illustrated and other methods are illustrated in this section.

Scatterplots similar to those produced with the `plot` function in Chapter 3 can be drawn using the `xyplot` function of the `lattice` package. Suppose we want to display the trend of the speed of Verlander's four-seam fastballs throughout the season, and contrast this trend across the four seasons. As a first step, the `subset` function is used to create a new data frame `F4verl` containing only Verlander's four-seam fastballs. Using the `format` and `as.integer` functions, the `gamedate` variable is converted to a variable `gameDay` containing the day of the year as an integer from 1 to 365. Using the `aggregate` function, the average speeds by day of the year and season are computed and stored in the variable `dailySpeed`.

```
F4verl <- subset(verlander, pitch_type == "FF")
F4verl$gameDay <- as.integer(format(F4verl$gamedate,
format="%j"))
```

```
dailySpeed <- aggregate(speed ~ gameDay + season, data=F4ver1,
   FUN=mean)
```

The `xyplot` function is used to construct a multipanel conditioning scatterplot of `speed` against `gameDay`. The syntax `speed ~ gameDay` indicates the scatterplot, and the vertical line (|) followed by the grouping variable `factor(season)` indicates that separate scatterplots are to be constructed for each season. The `xlab` and `ylab` arguments control the labels appearing on the axes.

```
xyplot(speed ~ gameDay | factor(season),
       data=dailySpeed,
       xlab="day of the year",
       ylab="pitch speed (mph)")
```

There are interesting patterns in these scatterplots. For example, Verlander's fastballs seem to get faster later in the 2012 season, and some of Verlander's slowest pitches in the 2010 season occurred towards the end of the season.

To illustrate a different graph, suppose we are interested in comparing Verlander's fastball and change-up speeds in this four-year period. The `verlander` data frame is restricted to only include the pitches labeled as either four-seam fastballs or change-ups. Then the `aggregate` function is used to calculate the average speeds by season and pitch type. (The `droplevels` function is applied to get rid of the unused factor levels.)

```
speedFC <- subset(verlander, pitch_type %in% c("FF", "CH"))
avgspeedFC <- aggregate(speed ~ pitch_type + season,
   data=speedFC, FUN=mean)
avgspeedFC <- droplevels(avgspeedFC)
avgspeedFC
```

	pitch_type	season	speed
1	CH	2009	85.06900
2	FF	2009	96.46576
3	CH	2010	86.72249
4	FF	2010	96.23772
5	CH	2011	87.56312
6	FF	2011	95.83171
7	CH	2012	87.38355
8	FF	2012	95.46240

The `dotplot` function in the `lattice` package is used to construct a dot plot to compare the fastball/change-up speed differential across seasons. By use of the `pch` argument, the pitch type first letter is used as the plotting symbol and `cex` is used for controlling the symbol size.

```
dotplot(factor(season) ~ speed, groups=pitch_type,
        data=avgspeedFC,
        pch=c("C", "F"), cex=2)
```

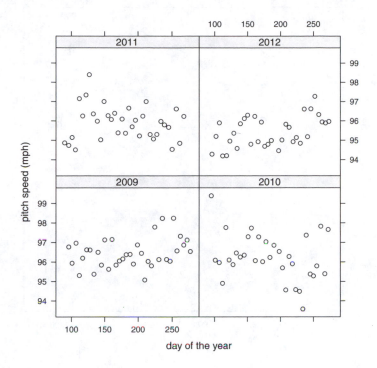

FIGURE 6.4

Scatterplots of the speeds of Verlander's four-seam fastball against the day of the year for the four seasons 2009 through 2012.

The graph of Figure 6.5 shows that Verlander's fastball/change-up speed differential has slightly diminished through the years, as Justin lost a very small amount of speed on his fastball, while simultaneously throwing his change-up progressively harder from 2009 to 2011.

6.2.7 The panel function

Is Verlander able to maintain the speed of his fastball throughout a game? The xyplot function in the lattice package is used to construct a scatterplot of pitch speed against the pitch count. This example illustrates the use of the panel function argument in lattice to overlay a vertical reference line at the 100 pitches mark and a horizontal reference line at the four-year average speed, and add some text and arrows to the graph.

R code to display the basic version of the function xyplot constructs a scatterplot of speed (pitch speed) against pitches (pitch count) is shown where the dataset (avgSpeed) is obtained by calculating the average of fastball

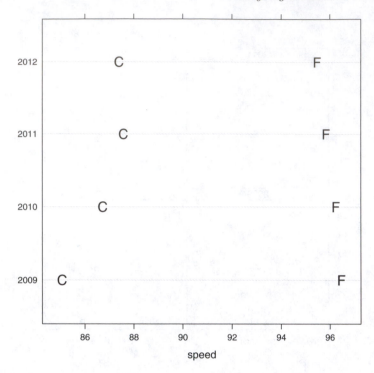

FIGURE 6.5
Dot plot of Verlander fastball and change-up speeds for the 2009 through 2012 seasons.

speeds at every pitch count for the four seasons considered. The overall average fastball speed is stored in the `avgSpeedComb` object.

```
avgSpeed <- aggregate(speed ~ pitches + season, data=F4ver1,
  FUN=mean)
xyplot(speed ~ pitches | factor(season),
      data=avgSpeed)
avgSpeedComb <- mean(F4ver1$speed)
```

To add the reference lines and text, one adds a `panel` argument which is a function describing the different components graphed. The particular `panel` function used for our example is displayed below. The `panel.xyplot` function draws the scatterplot, `panel.abline` functions add the vertical line at the 100 pitch count and the horizontal one at the average speed value, the `panel.text` function adds the text, and the `panel.arrows` function draws the arrows.

```
panel=function(...){
        panel.xyplot(...)
        panel.abline(v=100, lty="dotted")
```

```
          panel.abline(h=avgSpeedComb)
          panel.text(25, 100, "avg. speed")
          panel.arrows(25, 99.5, 0, avgSpeedComb,
                    length .1)
     }
```

The `panel.abline` and `panel.text` are the `lattice` equivalent functions to the `abline` and `text` functions in the base graphics described in Chapter 3. Similarly the `lattice` function `panel.arrows` corresponds to the `arrows` function in base graphics.[3]

If the `panel` function argument is added to the basic `xyplot` function, a scatterplot is obtained with the reference lines and text added (see Figure 6.6).

```
xyplot(speed ~ pitches | factor(season),
        data=avgSpeed,
        panel=function(...){
        panel.xyplot(...)
        panel.abline(v=100, lty="dotted")
        panel.abline(h=avgSpeedComb)
        panel.text(25, 100, "avg. speed")
        panel.arrows(25, 99.5, 0, avgSpeedComb,
                    length=.1)
     }
     )
```

This figure clearly shows that the speed of Verlander's fastball steadily increases during a game, even past the 100-pitch count, a commonly used limit for starting pitchers.

6.2.8 Building a graph, step-by-step

In this section a graph is built, step-by-step using the `lattice` package, using the material introduced in the previous sections of this chapter. For Justin Verlander's second no-hitter of his career (Tigers - Blue Jays game of May 7, 2011), we are interested in plotting the location of all of his pitches, taking in account the batter handedness and the pitch type. The data frame for the desired game `NoHit` is obtained using the `subset` function and the original data frame `verlander`.

```
NoHit <- subset(verlander, gamedate == "2011-05-07")
```

The variables `px` and `pz` give the horizontal and vertical pitch locations, the variable `batter_hand` contains the batter handedness and the variable

---

[3]The reader will find useful referring to the help entries for the base functions (for example, by typing `?arrows` in the R console) for figuring out the arguments that can be specified to the various panel functions.

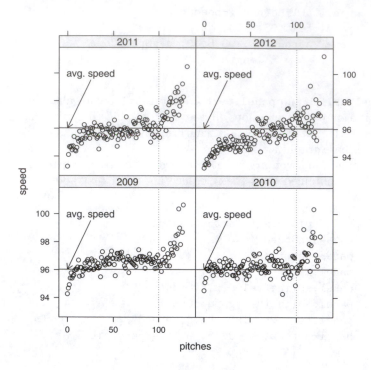

FIGURE 6.6
Verlander's four-seam fastball speed through the game - Scatterplot with vertical reference line at 100 pitches and horizontal reference line drawn at the value corresponding overall fastball speed (as indicated by the text labels and the arrows).

`pitch_type` contains the pitch type. A scatterplot of the pitch locations is constructed using the function `xyplot`; by use of the "`| batter_hand`" option, separate panels are created for left-handed and right-handed batters, and by use of the `groups=pitch_type` argument, each of the pitch types is represented by a different symbol. The `auto.key=TRUE` argument creates a legend for the plotting symbols. This first graph is displayed in Figure 6.7.

```
xyplot(pz ~ px | batter_hand, data=NoHit, groups=pitch_type,
       auto.key=TRUE)
```

Since the axes are both measured in feet, it is desirable that the units on the x-axis are expressed in the same scale of the units on the y-axis. This can be ensured by choosing isometric scales, which is accomplished by the `aspect="iso"` argument option. With this change, one obtains the graph shown in Figure 6.8.

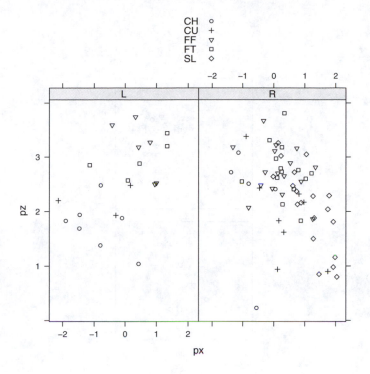

FIGURE 6.7
Location of Verlander's pitches in his second career no-hitter - base graph.

```
xyplot(pz ~ px | batter_hand, data=NoHit, groups=pitch_type,
       auto.key=TRUE,
       aspect="iso")
```

Next, we want to limit the plotting area from two feet to the left (of the center of the plate) to two feet to the right, and from the ground to five feet above it. The limits of the x and y axes are controlled by the arguments `xlim` and `ylim`. The axes can be labeled with meaningful text using the `xlab` and `ylab` arguments.[4] The third graph in our sequence is displayed in Figure 6.9.

```
xyplot(pz ~ px | batter_hand, data=NoHit, groups=pitch_type,
       auto.key=TRUE,
       aspect="iso",
       xlim=c(-2.2, 2.2),
       ylim=c(0, 5),
       xlab="Horizontal Location\n(ft. from middle of plate)",
       ylab="Vertical Location\n(ft. from ground)")
```

---

[4]The \n character sequence is used to split the text over multiple lines.

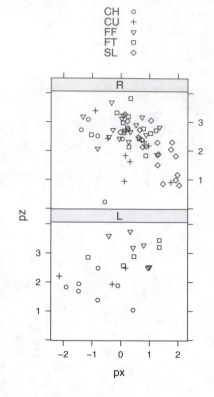

FIGURE 6.8
Location of Verlander's pitches in his second career no-hitter - graph with change in aspect ratio.

Further improvements can be made to the display in Figure 6.9. This graph is missing the rectangle of the strike zone and the legend uses abbreviations of the pitch type and has neither the title ("pitch type") nor the border.

The first change is to improve the legend. Other than a logical argument, the `auto.key` argument can accept a list of parameters to fine-tune the legend. The following code prepares this `list`, after a vector of labels for the pitch type is created.

```
pitchnames <- c("change-up", "curveball", "4S-fastball"
                , "2S-fastball", "slider")
myKey <- list(space="right",
               border=TRUE,
               cex.title=.8,
               title="pitch type",
               text=pitchnames,
               padding.text=4)
```

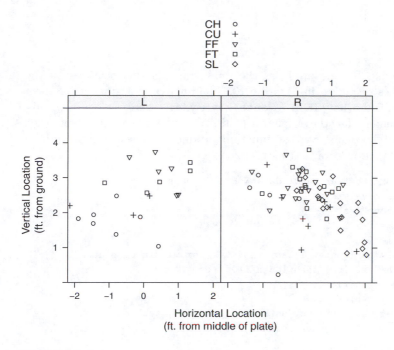

FIGURE 6.9

Location of Verlander's pitches in his second career no-hitter - graph with changes in axes limits and labels.

The meaning of most of the parameters should be clear. Of the others, `padding.text` indicates the spacing between the lines of the legend and `cex.title` is the font size for the `title` text.

To plot the strike zone on the figure, it is necessary to consult the baseball rule book on the size of this region. According to the rule book, the home plate defines the width of the strike zone. The front side of home plate measures 17 inches. Since the PITCHf/x coordinates system assumes the middle of the plate is located at the origin of the x-axis, the front side of the plate is located from $-17/12/2 \simeq -0.71$ to $17/12/2 \simeq 0.71$ feet. However, since just a piece of the baseball needs to cross home plate in order for the pitch to be considered a strike, the diameter of the ball (about 2.9 inches) has to be added to both sides, making the range $[-0.95; 0.95]$. The height of the strike zone varies from batter to batter and we chose to use average values for the top and the bottom of the zones (3.5 and 1.6 feet, respectively.) Based on this discussion, the limits for the strike zone rectangle can be defined in R by the assignments to the variables `topKzone`, `botKzone`, `inKzone`, and `outKzone`.

```
topKzone <- 3.5
botKzone <- 1.6
inKzone <- -.95
outKzone <- 0.95
```

The following R code produces the final graph of the locations of Verlander's pitches displayed in Figure 6.10. The `auto.key` argument has been changed from `auto.key=TRUE` to `auto.key=myKey` where `myKey` is the list of our parameters for the legend. In addition, the `panel` function argument is new. The `panel.xyplot` function produces the scatterplot and the `panel.rect` function[5] draws the strike zone.

```
xyplot(pz ~ px | batter_hand, data=NoHit, groups=pitch_type,
       auto.key=myKey,
       aspect="iso",
       xlim=c(-2.2, 2.2),
       ylim=c(0, 5),
       xlab="horizontal location\n(ft. from middle of plate)",
       ylab="vertical location\n(ft. from ground)",
       panel=function(...){
       panel.xyplot(...)
       panel.rect(inKzone, botKzone, outKzone, topKzone,
                  border="black", lty=3)
    }
  )
```

This final graph is very informative about the location of Verlander's pitches during this no-hit game. For example, we see that Verlander was successful in throwing his slider down and away to right-handed hitters. We also see Verlander's tendency to throw his change-up down and away to left-handed hitters. His fastballs generally were thrown within the strike zone.

6.3   The ggplot2 Package

6.3.1   Introduction

The `ggplot2` graphics system developed by Hadley Wickham is an implementation in R of Leland Wilkinson's framework described in Wilkinson (2005). Wilkinson describes a statistical graphic as a combination of independent components. Variables in a dataset are assigned particular roles or aesthetics; the aesthetic for one variable might be the plotting position along the horizontal axis and the aesthetic for a second variable might be the color or shape of the plotting point. Once the aesthetics for a set of variables are defined, then one

---

[5]Equivalent to the `rect` function for base graphics.

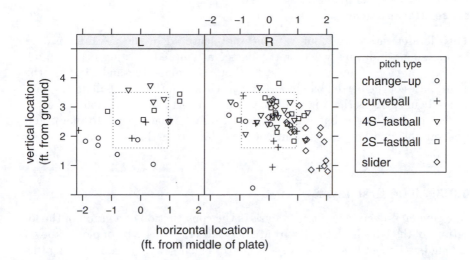

FIGURE 6.10

Location of Verlander's pitches in his second career no-hitter - graph with changes in legend and addition of strike zone box.

uses a geometric object or geom to construct a graph. Examples of geoms are points, lines, bars, histograms, and boxplots.

Graphics in `ggplot2` are constructed by progressively adding layers, starting with the raw data, then adding strata of statistical summaries and annotations. This section illustrates the layer-by-layer building of a `ggplot2` graphic.

6.3.2 The `cabrera` dataset

To illustrate `ggplot2` graphics, we consider the `cabrera` dataset, featuring PITCHf/x batting data for five seasons of Miguel Cabrera's career including the historical 2012 season in which he was the first winner of the batting Triple

Crown since Carl Yastrzemski's accomplishment in 1967. A random sample of twenty rows of the dataset is displayed in Table 6.2 using the following code.

```
sampleRows <- sample(1:nrow(cabrera), 20)
cabrera[sampleRows,]
```

This dataset introduces four variables not featured in the `verlander` data frame: `swung` indicates whether the batter attempted a swing on the pitch (coded 1 in case of an attempt and 0 otherwise), and `hitx` and `hity` are the coordinates of the batted ball in feet (for pitches resulting in a ball hit in-play) where home plate is the origin. Last, the variable `hit_outcome` is an indicator of the result of the at-bat (again reported only on balls put into play), and this variable can be either `O` (out), `H` (base hit), or `E` (batter reaching on an error).

6.3.3   The first layer

One general objective in pitch analysis is the construction of a graph of the locations of the balls hit into play by Miguel Cabrera – this type of plot is known among baseball enthusiasts as a *spray chart*. The first step in constructing this graph in the `ggplot2` package is to find the relevant data frame and assign roles to the scatterplot variables. The `ggplot2` package is first loaded by the `library` function. In the `ggplot` function, the `data` argument indicates the use of the `cabrera` data frame and the `aes` argument indicates that `hitx` (the horizontal coordinate of the batted ball) is assigned the position on the x-axis and `hity` (the vertical coordinate of the batted ball) is assigned the position on the y-axis.

```
library(ggplot2)
p0 <- ggplot(data=cabrera, aes(x=hitx, y=hity))
```

The `ggplot` object is assigned to the variable p0, allowing for the addition of the ensuing layers. Typing p0 in the R console at this point does not produce a plot as we have just set up the defaults for the plot and no plotting layer has been yet created.[6]

To produce a graph, one needs to add at least one layer to the `ggplot` object. For example, the `geom_point` function adds a points layer and creates a scatterplot. One sees the graph shown in Figure 6.11 by typing p1 in the R console.

```
p1 <- p0 + geom_point()
p1
```

No argument is passed to the `geom_point` function, indicating the defaults set in the previous `ggplot` call are to be maintained. One can override the set defaults for one or more layers, as demonstrated in Sections 6.3.6 and 6.3.8.

---

[6]In fact R would return the following error message: `Error: No layers in plot`.

TABLE 6.2

Twenty rows sample of the cabrera dataset.

(row#)	season	gamedate	pitch_type	balls	strikes	speed	px	pz	swung	hitx	hity	hit_outcome
1512	2009	2009-10-01	SL	0	1	80.50	-0.02	2.94	1			
1866	2010	2010-05-09	CU	1	1	77.90	0.47	2.66	1	78.60	169.60	O
2776	2010	2010-08-31	FF	0	0	90.80	-2.46	1.82	0			
2645	2010	2010-08-18	FC	0	1	85.40	0.77	2.80	1			
5361	2012	2012-07-04	SL	0	0	85.50	-1.25	2.33	0			
5227	2012	2012-07-15	SL	0	2	83.60	2.14	1.92	0			
4285	2011	2011-09-02	CH	1	1	83.30	0.49	3.05	1	94.80	56.19	O
1759	2010	2010-04-25	FF	0	1	87.90	-0.40	2.73	1	-32.15	123.71	O
5048	2012	2012-08-03	SL	0	0	83.90	2.69	1.69	0			
5186	2012	2012-07-13	SL	3	2	85.40	-0.43	1.93	1	32.68	339.77	O
5091	2012	2012-08-06	FF	2	0	95.10	0.24	2.02	0			
5655	2012	2012-06-07	SI	3	0	89.00	0.46	1.99	0			
2510	2010	2010-08-03	FT	3	1	86.00	1.01	3.33	1			
3939	2011	2011-07-28	SI	0	0	93.90	0.01	2.58	0			
2185	2010	2010-06-23	KN	0	0	78.70	-0.41	2.05	1	-48.34	77.79	O
2410	2010	2010-07-21	FF	0	0	92.40	-1.56	1.72	0			
5427	2012	2012-06-22	FF	1	2	95.90	-0.20	1.84	1	13.77	37.28	O
2897	2010	2010-09-18	FF	0	1	96.50	-0.24	3.76	1			
5187	2012	2012-07-13	FF	2	2	94.50	-1.72	3.36	0			
5477	2012	2012-06-12	SL	3	2	83.10	-0.46	1.98	1	-61.84	77.79	O

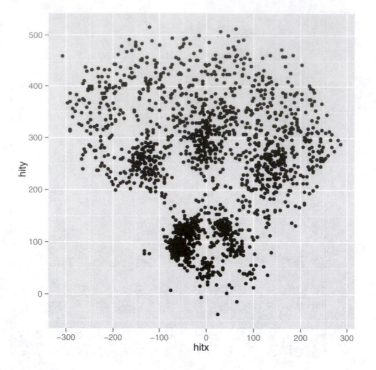

FIGURE 6.11
Scatterplot of Miguel Cabrera's batted balls (2009 - 2012).

6.3.4   Grouping factors

The `hit_outcome` variable in the `cabrera` data frame indicates whether a batted ball resulted in a base hit (`H`), an out (`O`), or an error (`E`). By assigning the variable `hit_outcome` to the `color` aesthetic in the `ggplot` function, the three different outcomes are represented with different colors in the plot. The `coord_equal` function ensures the units are equally scaled on the x-axis and on the y-axis (as they are measured in feet in both cases); it is the equivalent of passing the `aspect="iso"` argument to a `lattice` plot. The plot (Figure 6.12) is viewed after typing p2 in the R console.

```
p0 <- ggplot(data=cabrera, aes(hitx, hity))
p1 <- p0 + geom_point(aes(color=hit_outcome))
p2 <- p1 + coord_equal()
p2
```

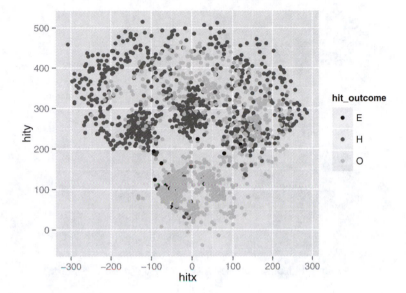

FIGURE 6.12
Outcomes of Miguel Cabrera's batted balls (2009 - 2012).

6.3.5 Multipanel conditioning (faceting)

Similar to `lattice` graphics, multipanel conditioning plots are straightforward to construct in `ggplot2` graphics. The different panels in which one subsets data are described in `ggplot2` as *facets*. A multipanel conditioning plot is constructed with `season` as the conditioning variable by adding the `facet_wrap` function with the ∼ `season` argument to the current plot `p2`. (See Figure 6.13.)

```
p3 <- p2 + facet_wrap(~ season)
p3
```

This figure illustrates how the locations of Cabrera's batted balls have changed over the four seasons.

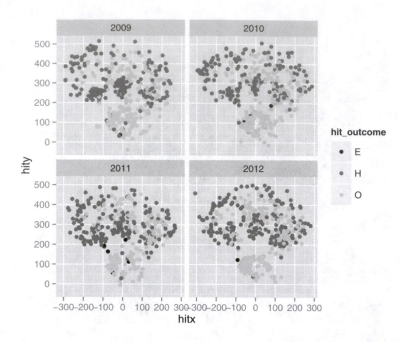

FIGURE 6.13
Outcomes of Miguel Cabrera's batted balls by season.

6.3.6   Adding elements

One can add reference lines in `ggplot2` as in other graphics systems. For
example, one may want to display the base paths in the previous plots. Since
home plate is set to be at the origin and feet are the unit of measurement
in the [`hitx`, `hity`] coordinates system, the positions of the remaining three
bases are obtained by a few applications of the Pythagorean theorem.[7] A new
data frame `bases` is defined that gives the location of the bases in (x, y)
coordinates.

```
bases <- data.frame(x=c(0, 90/sqrt(2), 0, -90/sqrt(2), 0),
              y=c(0, 90/sqrt(2), 2 * 90/sqrt(2), 90/sqrt(2), 0)
                  )
```

---

[7]The distance between bases as defined by the baseball rule book is 90 feet.

Note that, since we are going to draw a path, the coordinates of home plate are contained both at the beginning and at the end of the values of x and y. The displaying of the base paths lines (Figure 6.14) is obtained by adding a layer to the previous plot.

The new plot with base paths added is constructed by adding the geom_path function component to the current plot p3. In this case, data and the mapping of aesthetics (aes) are specified as arguments in the geom_path function, overriding the defaults set in the original ggplot function. When calling the p4 object, two additional layers are added for the display of the foul lines using the geom_segment functions.

```
p4 <- p3 + geom_path(aes(x=x, y=y), data=bases)
p4 +
  geom_segment(x=0, xend=300, y=0, yend=300) +
  geom_segment(x=0, xend=-300, y=0, yend=300)
```

6.3.7   Combining information

Using the ggplot2 graphics system it is easy to combine multiple information on a single plot. For example, the graph in Figure 6.15 is an enhanced display of Cabrera's regular season batted ball data from September and October 2012. This plot simultaneously portrays the batted ball outcome (mapped to the point shape), the pitch speed (mapped to the point size), and the pitch type (mapped to the point color). This graph is constructed by use of the ggplot function with the variable hit_outcome assigned the shape aesthetic, the variable pitch_type assigned the color aesthetic, and speed assigned the size aesthetic. The guides function allows control on how the legend is placed – in this case, the col=guide_legend(ncol=2) argument indicates the color key is distributed over two columns.

```
cabreraStretch <- subset(cabrera, gamedate > "2012-08-31")
p0 <- ggplot(data=cabreraStretch, aes(hitx, hity))
p1 <- p0 + geom_point(aes(shape=hit_outcome, colour=pitch_type,
                          size=speed))
p2 <- p1 + coord_equal()
p3 <- p2 + geom_path(aes(x=x, y=y), data=bases)
p4 <- p3 + guides(col=guide_legend(ncol=2))
p4 +
  geom_segment(x=0, xend=300, y=0, yend=300) +
  geom_segment(x=0, xend=-300, y=0, yend=300)
```

6.3.8   Adding a smooth line with error bands

We return to the earlier scatterplot of the pitch speed against pitch count of Justin Verlander discussed in Section 6.2.7. Using ggplot2, one can create a figure analogous to Figure 6.6, with the added bonus of also provid-

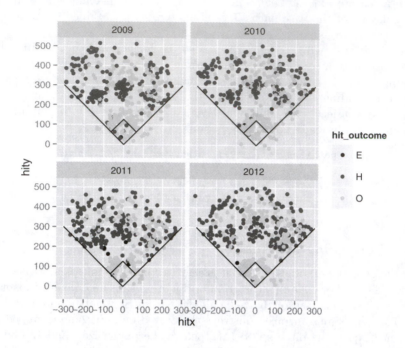

FIGURE 6.14
Outcomes of Miguel Cabrera's batted balls by season (with base lines super-posed).

ing smoothed lines with shading reflecting the error bands (Figure 6.16).[8]
This graph is constructed in ggplot2 in five layers. The geom_point function
gives the plotted points, the facet_wrap produces separate panels by year,
the geom_smooth function provides the smoothed line with error bands, the
geom_vline function gives the vertical line at 100 pitches, and the geom_line
function draws the horizontal line.[9]

```
ggplot(F4ver1, aes(pitches, speed)) +
  facet_wrap(~ season) +
  geom_line(stat="hline", yintercept="mean", lty=3) +
  geom_point(aes(pitches, speed),
```

---

[8]Adding smoothed lines to the lattice version of the plot would also be possible, but
would require the use of much more complicated code.
[9]Note that in this instance we chose to display in each panel the average speed for the
season displayed in the panel.

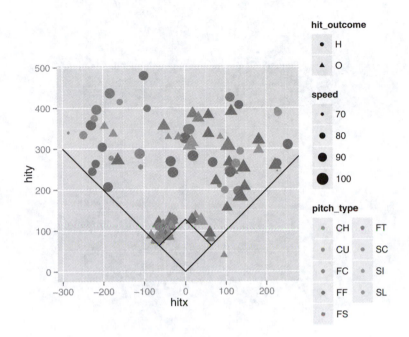

FIGURE 6.15

Miguel Cabrera's batted balls in the final month of his Triple Crown season (September/October 2012).

```
             data=F4verl[sample(1:nrow(F4verl), 1000),]) +
    geom_smooth(col="black") +
    geom_vline(aes(xintercept=100), col="black", lty=2)
```

To avoid an over-cluttering of data points in the graph, the aesthetics mapping is respecified in the geom_point layer, where a random sample of 1000 points is used from the F4verl data frame. The smoothing has been calculated on the original F4verl data frame, because no explicit aesthetics mapping has been specified in the geom_smooth layer.

6.3.9 Dealing with cluttered charts

In our example, plots have been built in ggplot2 by adding geom layers. However, this graphics system also allows the addition of stat layers, consisting

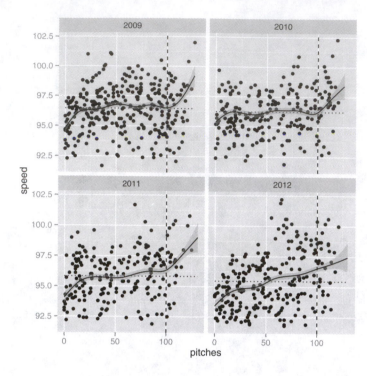

FIGURE 6.16
Verlander's four-seam fastball speed through the game - Scatterplot with reference line at 100 pitches and a smooth line with shading for errors.

of statistical transformations of the data. For example, one can add summary statistics such as means and standard deviations to the current graph.

These `stat` layers are useful for dealing with cluttered data. As an example, if we are to use a scatterplot to graph the locations of Verlander's four-seam fastballs from 2009 to 2012 using the following code, we see an indistinguishable cloud of black points and it is difficult to see any patterns (Figure 6.17).

```
kZone <- data.frame(
  x=c(inKzone, inKzone, outKzone, outKzone, inKzone),
  y=c(botKzone, topKzone, topKzone, botKzone, botKzone)
  )
ggplot(F4verl, aes(px, pz)) +
  geom_point() +
  facet_wrap(~ batter_hand) +
  coord_equal() +
  geom_path(aes(x, y), data=kZone, lwd=2, col="white")
```

One way of handling the cluttering of data points is tiling the $[x; y]$ plane

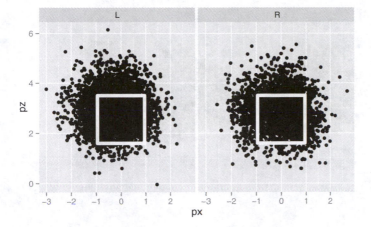

FIGURE 6.17
A cluttered scatterplot - Locations of Justin Verlander's four-seam fastballs by batter handedness (2009-2012).

with hexagonal bins and having the bins colored according to the number of data points they contain. One can create a new graph by replacing the `geom_point` layer with a `stat_binhex` layer. The new graph is displayed in Figure 6.18.

```
ggplot(F4verl, aes(px, pz)) +
  stat_binhex() +
  facet_wrap(~ batter_hand) +
  coord_equal() +
  geom_path(aes(x, y), data=kZone, lwd=2, col="white", alpha=.3)
```

Note how the `alpha` argument in the `geom_path` layer is used to adjust the transparency of the strike zone border.

6.3.10 Adding a background image

Adding images to the plot background is usually not a good idea, as they often do nothing more than obscure the data points. However, plotting batted

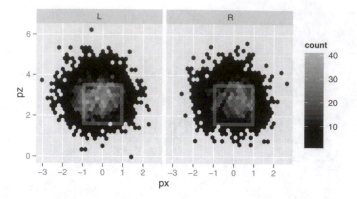

FIGURE 6.18
Hexagonal binning as a way to portray 2D density - Locations of Justin Verlander's four-seam fastballs by batter handedness (2009-2012).

ball locations might be an exception, since the figure of the baseball field may give a better reference guide than the horizontal and vertical axes.

Packages exist for reading common format images into R. For example, if one obtains a diagram of Detroit's Comerica Park[10] as a *jpeg* file, the `jpeg` package can be used for reading the image into R by the `readJPEG` function.[11]

```
library(jpeg)
diamond <- readJPEG("Comerica.jpg")
```

The `diamond` object is a three-dimensional array of dimension $x \times y \times 3$, where x and y correspond to the dimension of the image in pixels. The array can be seen as a collection of three $x \times y$ matrices, containing information on the red, green, and blue (RGB) components at every pixel of the image respectively, expressed as values in the $[0, 1]$ range.

---

[10]This diagram was retrieved as a *svg* file from MLBAM Gameday at `gd2.mlb.com/images/gameday/fields/svg/2394.svg` and then converted to a *jpeg* format.
[11]Similar packages named `png` and `bmp` provide analogous functionality for reading other common image formats.

The diagram can be added in `ggplot2` as a layer of the plot using the `annotation_raster` function. The following code is used to obtain Figure 6.19.

```
ggplot(cabrera, aes(hitx, hity)) +
  coord_equal() +
  annotation_raster(diamond, -310, 305, -100, 480) +
  stat_binhex(alpha=.9, binwidth=c(5, 5)) +
  scale_fill_gradient(low="grey70", high="black")
```

The four numbers passed to the `annotation_raster` layer indicate where the image has to be positioned and were found by trial and error.[12] In the `stat_binhex` layer two arguments have been specified: `alpha` gives the degree of transparency to the hexagons so that their coloring does not completely hide the diamond diagram and `binwidth` sets the dimensions of the hexagons. Finally, the `scale_fill_gradient` layer sets the coloring of the hexagonal bins as a gradient starting from the color `grey70` and ending at the color `black`.[13]

6.4 Further Reading

Detailed descriptions of the `lattice` and `ggplot2` packages are available in Sarkar (2008) and Wickham (2009) by the respective developers.

6.5 Exercises

1. (**Location of Pitches for Left- and Right-Handed Batters**)

   Use a density plot to display the horizontal location of Justin Verlander's pitchers by opponent's handedness. Choose the conditioning and grouping variables so that one can easily detect the differences in location by handedness. Add a legend (if necessary) and vertical reference lines indicating the borders of the strike zone.

2. (**Comparing Pitch Locations for Two Pitchers**)

---

[12]Since the image is imported in R as an array of RGB values, it would be possible to retrieve the position of two points (for example home plate and second base) and compute the four values to be passed in the `annotation_raster` layer. For an example on how to identify an object inside an image with R look at `is-r.tumblr.com/post/36874307174/finding-a-bright-object`.

[13]A map of R colors by name is available at `research.stowers-institute.org/efg/R/Color/Chart/ColorChart.pdf`.

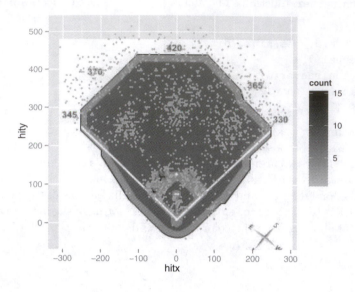

FIGURE 6.19
Scatterplot of Cabrera's batted balls (2009 - 2012) with Detroit's Comerica
Park diagram in the background. Note: batted balls for games in other ball-
parks are included as well.

The `sanchez` data frame contains 2008-2012 PITCHf/x data for pitcher
Jonathan Sanchez. The structure of this data frame is similar to the
`verlander` data frame described in the chapter. Use a graphical display
to compare the ability of Sanchez and Verlander in maintaining their fastball speed through the game. (See Sections 6.2.7 and 6.3.8.) Use either
the `lattice` or `ggplot2` graphics package and display the data either as
a multipanel plot or a superposed lines plot.

3. **(Graphical View of the Speeds of Justin Verlander's Fastballs)**

   (a) The `cut` function is useful for recoding a continuous variable into
       intervals. Use this function to categorize the `pitches` variable in the
       `verlander` data frame in groups of ten pitches.

   (b) Use the `bwplot` function from the `lattice` package to produce a
       boxplot of Verlander's four-seam fastball speed (use the `F4verl` data
       frame) for each ten-pitches group. Compare the information conveyed
       by the resulting chart with that of Figure 6.6.

4. (**Exploring Miguel Cabrera's Slugging Power**)

    (a) Create a data frame by selecting, from the `cabrera` data frame, the instances where the `hit_outcome` variable assumes the value H (for base hit).

    (b) Using the `hitx` and `hity` variables, create a new variable equal to the distance, in feet from home plate, of Cabrera's base hits. (This variable is computed by simply applying the Pythagorean Theorem– remember that home plate is at the origin.)

    (c) In the newly created data frame, create a `gameDay` variable indicating the day of the year (from 0 to 365) in which the game took place (see Section 6.2.6).

    (d) Build a scatterplot featuring `gameDay` on the x-axis, `distance` on the y-axis and a smooth line with error bands. Does the resulting plot appear to indicate changes in Cabrera's power during the season?

7

Balls and Strikes Effects

CONTENTS

7.1	Introduction		161
7.2	Hitter's Counts and Pitcher's Counts		162
	7.2.1	Introduction	162
	7.2.2	An example for a single pitcher	162
	7.2.3	Pitch sequences on Retrosheet	165
		7.2.3.1 Functions for string manipulation	165
		7.2.3.2 Finding plate appearances going through a given count	167
	7.2.4	Expected run value by count	169
	7.2.5	The importance of the previous count	170
7.3	Behaviors by Count		173
	7.3.1	Swinging tendencies by count	173
		7.3.1.1 Propensity to swing by location	173
		7.3.1.2 Effect of the ball/strike count	176
	7.3.2	Pitch selection by count	178
	7.3.3	Umpires' behavior by count	181
7.4	Further Reading		184
7.5	Exercises		185

7.1 Introduction

In this chapter we explore the effect of the ball/strike count on the behavior of players and umpires and on the final outcome of a plate appearance. Retrosheet data from the 2011 season is used to estimate how the ball/strike count affects the run expectancy. Also PITCHf/x data is used to explore how one pitcher modifies his pitch selection based on the count and, similarly, how one batter alters his swing zone, and umpires judge pitches according to the count. Functions for string manipulation are introduced that are useful for managing the pitch sequences from the Retrosheet files. Level plots and contour plots, created with the use of the `lattice` package, will be used for the explorations of batters' swing tendencies and umpires' strike zones.

7.2   Hitter's Counts and Pitcher's Counts

7.2.1   Introduction

When watching a broadcast of a baseball game, one often hears an announcer's concern for a pitcher who is constantly "falling behind" in the count, or his/her anticipation for a particular pitch because it's a "hitter's count" and the batter has a chance to do some damage. We will see if there is actual evidence that the so-called hitter's count really leads to more favorable outcomes for batters, while "getting ahead" in the count (a pitcher's count) is beneficial for pitchers.

7.2.2   An example for a single pitcher

The Baseball Reference[1] website provides for every player a plethora of splits statistics. The website provides splits by ball/strike counts for all seasons since 1988. Mike Mussina's split statistics are found by entering the player's profile page (typing "Mussina" on the search box brings one there), clicking on the "Splits" tab in the "Standard Pitching" table, and clicking on "Career" (or whatever season we are interested in) on the pop-up menu that appears. One finds the "Count Balls/Strikes" table scrolling down on the splits page. Alternatively, the table can be reached by a direct link: in this case the career splits by count for Mike Mussina are currently available at `www.baseball-reference.com/players/split.cgi?id= mussimi01&year=Career&t=p#count`.

The first series of lines (from "First Pitch" to "Full Count") shows the statistics for events happening in that particular count. Thus, for example, a batting average (BA) of .337 on 1-0 counts indicates batters hit safely 34% of the time when putting the ball in play *on a 1-0 count* against Mussina. We are more interested in the second group of rows, those beginning with the word "After". In fact, in these cases the statistics are relative to every plate appearance that *goes through* that count. Thus a .337 on-base percentage (OBP) after 1-0 means that, whenever Mike Mussina started a batter with a ball, the batter successfully got on base 34% of the time, no matter how many pitches it took to end the plate appearance.

The last column on every table in the splits page is "tOPS+". It's an index for comparing the overall player's OPS (the sum of on base percentage and slugging percentage[2]) with his OPS in that particular situation. A value over 100 is interpreted as a higher OPS in the situation compared to the overall OPS; conversely values below 100 indicate a OPS value that is lower than the overall OPS.

---

[1]`baseball-reference.com`

[2]OPS is widely used as a measure of offensive production because, while being very easy to calculate, it correlates very well, at the team level, with runs scored.

Figure 7.1 uses a heat map to display Mussina's tOPS+ through the various counts. If one focuses on a particular number of strikes, a higher number of balls in the count makes the outcome more likely to be favorable to the hitter (lighter shades). Conversely, if one fixes the number of balls, the balance moves towards the pitcher (darker shades) as one increases the number of strikes. This figure emphasizes the importance from a pitcher's perspec-

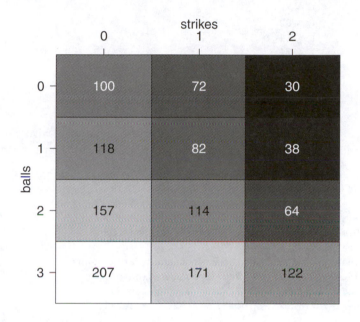

FIGURE 7.1
Heat map of tOPS+ for Mike Mussina through each balls/strikes count. Data from Baseball Reference website.

tive of beginning the duel with a strike. When Mussina fell behind 1-0 in his career, batters performed 18% better than usual in the plate appearance; conversely, after a first pitch strike, they were limited to 72% of their potential performance.

How is the heatmap display of Figure 7.1 created in R? First a data frame `mussina` is prepared with all the possible balls/strikes counts, using the `expand.grid` function as previously illustrated in Section 4.7. A new variable `value` is added with the tOPS+ values taken from the Baseball-Reference website.

```
mussina <- expand.grid(balls=0 : 3, strikes=0 : 2)
mussina$value <- c(100, 118, 157, 207, 72, 82, 114, 171, 30, 38,
```

```
   64, 122)
mussina
```

```
   balls strikes value
1      0       0   100
2      1       0   118
3      2       0   157
4      3       0   207
5      0       1    72
6      1       1    82
7      2       1   114
8      3       1   171
9      0       2    30
10     1       2    38
11     2       2    64
12     3       2   122
```

The graph is created using the new function `countmap` with argument `mussina`, making use of the `color2D.matplot` function in the `plotrix` package.

```
countmap <- function(data){
  require(plotrix)
  data <- xtabs(value ~ ., data)
  color2D.matplot(data, show.values=2, axes=FALSE
    , xlab="", ylab="")
  axis(side=2, at=3.5:0.5, labels=rownames(data), las=1)
  axis(side=3, at=0.5:2.5, labels=colnames(data))
  mtext(text="balls", side=2, line=2, cex.lab=1)
  mtext(text="strikes", side=3, line=2, cex.lab=1)
}
countmap(mussina)
```

Let's explore the commands inside the `countmap` function. The first line loads the package `plotrix`. (This package needs to be installed beforehand for the function to work.) The second line reshapes the data in a contingency table format that is needed for creating the heatmap table. For example, passing the `mussina` data frame as the `data` argument, the data frame looks like the following after applying the `xtabs` function:

```
      strikes
balls   0    1    2
    0 100   72   30
    1 118   82   38
    2 157  114   64
    3 207  171  122
```

The `color2D.matplot` function is called to draw the plot. By use of the argument `show.values=2`, we indicate that the values (tOPS+) are to be shown.[3]

---

[3]The parameter `show.values` requires either a logical value (TRUE/FALSE), indicating

The arguments `axes=FALSE, xlab="", ylab=""` indicate that axes and labels are not drawn. The axes and labels are explicitly drawn by use of the `axis` and `mtext` functions.

The `axis` function controls how axes are drawn. The side of the plot where the axis should appear is controlled by the `side` parameter; to place an axis on top of the plot, as was done in Figure 7.1, one specifies `side=3`. The `at` parameter controls the positions at which tick marks are to be drawn, while `labels` indicates the text that is to be placed at the tick points. Several other optional parameters can be passed to the `axis` function to control the positioning of the axis, the appearance of the line, and the tick marks and the font for the labels (type `?axis` on the R console for more details).

The `mtext` function allows to place text at the margins of a plot. Similarly to what happens for `axis`, the `side` parameter specifies whether the text should be placed to the bottom (`side=1`), the left (2), the top (3), or the right (4) of the plot. Other parameters are used to indicate the distance from the margin at which the text should be placed (`line`) and its appearance.

7.2.3 Pitch sequences on Retrosheet

From viewing Figure 7.1, one obtains an initial view of hitter's counts (lighter shades) and pitcher's counts (darker shades) on the basis of offensive production. However this figure is based on data for a single pitcher – does a similar pattern emerge when using league-wide data?

Retrosheet provides pitch sequences beginning with the 1988 season. Sequences are stored in strings such as the example `FBSX`. Each character encodes the description of a pitch. In this example, the pitch sequence is a foulball, followed by a ball, a swinging strike, and the ball put into play. Table 7.1 provides the description for every code used in Retrosheet pitch sequences.[4]

7.2.3.1 Functions for string manipulation

Sequence strings from Retrosheet need some initial processing to place in a form suitable for analysis. A quick tutorial on some R functions for the manipulation of strings is provided in this section. Readers not interested in string manipulation functions may skip to Section 7.2.4.

The function `nchar` returns the number of characters in a string. This function is helpful for obtaining the number of pitches delivered in a Retrosheet pitch sequence. For example, the number of pitches in the sequence `BBSBFFFX` is given by

```
nchar("BBSBFFFX")
[1] 8
```

whether the values should be displayed inside the cells, or an integer number, dictating the maximum number of decimal places to be displayed.

[4]Source: `www.retrosheet.org/eventfile.htm`.

TABLE 7.1

Pitch codes used by Retrosheet.

Symbol	description
+	following pickoff throw by the catcher
*	indicates the following pitch was blocked by the catcher
.	marker for play not involving the batter
1	pickoff throw to first
2	pickoff throw to second
3	pickoff throw to third
>	indicates a runner going on the pitch
B	ball
C	called strike
F	foul
H	hit batter
I	intentional ball
K	strike (unknown type)
L	foul bunt
M	missed bunt attempt
N	no pitch (on balks and interference calls)
O	foul tip on bunt
P	pitchout
Q	swinging on pitchout
R	foul ball on pitchout
S	swinging strike
T	foul tip
U	unknown or missed pitch
V	called ball because pitcher went to his mouth
X	ball put into play by batter
Y	ball put into play on pitchout

However, as indicated in Table 7.1, there are some characters in the Retrosheet strings denoting actions that are not pitches, such as pickoff attempts.

The functions `grep` and `grepl` are used to find a pattern within elements of character vectors. The function `grep` returns the indices of the elements for which a match is found, and the function `grepl` returns a logical vector, indicating for each element of the vector whether a match is found. For both functions, the first argument is the string pattern to search and the second argument is the vector of strings where matches are sought. For example, we apply the two functions to the vector of pitch sequences `sequences` in search for pickoff attempts to first base denoted by the code 1.

```
sequences <- c("BBX", "C11BBC1S", "1X")
grep("1", sequences)
[1] 2 3
grepl("1", sequences)
[1] FALSE  TRUE  TRUE
```

The function `grep` tells us that "1" is contained in the second (2) and third (3) components of the character vector `sequences`, and `grepl` outputs this same information by means of a logical vector.

The pattern parameter to search does have to be a single character. For example we may want to look for consecutive pickoff attempts to first which is the pattern "11". The below output shows that "11" is contained in the second component of `sequences`.

```
grepl("11", sequences)
[1] FALSE  TRUE FALSE
```

Also the function `gsub` allows for the substitution of the pattern found with a replacement. The replacement can be an empty string, in which case the pattern is simply removed. For example, the following code removes the pickoff attempts to first from the pitch sequences.

```
gsub("1", "", sequences)
[1] "BBX"    "CBBCS" "X"
```

7.2.3.2 Finding plate appearances going through a given count

Since we are interested only in pitch counts, it is necessary to remove the characters not corresponding to actual pitches from the pitch sequences. Regular expressions are the computing tool needed for this particular task. While it's beyond the scope of this book to fully explain how regular expressions work, we will instead show a few examples on how to use them.[5]

We begin by loading the `all2011.csv` file containing Retrosheet's play-by-play for the 2011 season.

---

[5]The website `www.regular-expressions.info/` is a very comprehensive online resource on regular expressions, featuring examples, tutorials, reference for syntax, and a list of related books.

```
pbp2011 <- read.csv("retrosheet/all2011.csv")
headers <- read.csv("retrosheet/fields.csv")
names(pbp2011) <- headers$Header
```

The `gsub` function is used to create the new variable `pseq` of pitch sequences which removes the symbols from the Retrosheet pitch sequence variable `PITCH_SEQ_TX` that don't correspond to actual pitches.

```
pbp2011$pseq <- gsub("[.>123N+*]", "", pbp2011$PITCH_SEQ_TX)
```

The square brackets indicate, in a regular expression, the collection of characters to search. The above code removes pickoff attempts at any base (1, 2, 3) either by the pitcher or the catcher (+), balks and interference calls (N), plays not involving the batter (.), indicators of runners going on the pitch (>), and of catchers blocking the pitch (*).[6]

Another special character is used to identify the plate appearances that go through a 1-0 count. In a regular expression, the ^ character means the pattern has to be matched at the beginning of the string. Looking at Table 7.1 for the four different ways a ball can be coded, the following line script creates the desired variable `c10`.

```
pbp2011$c10 <- grepl("^[BIPV]", pbp2011$pseq)
```

Similarly, plate appearances going through a 0-1 count are identified by use of the new variable `c01` variable.

```
pbp2011$c01 <- grepl("^[CFKLMOQRST]", pbp2011$pseq)
```

To check our work, the values of `PITCH_SEQ_TX`, `c10`, and `c01` are displayed for the first ten lines of the data frame.

```
pbp2011[1:10, c("PITCH_SEQ_TX", "c10", "c01")]
```

```
     PITCH_SEQ_TX    c10    c01
1               X  FALSE  FALSE
2            CBCS  FALSE   TRUE
3           CBBBB  FALSE   TRUE
4           BCSBS   TRUE  FALSE
5          CBB1>S  FALSE   TRUE
6      CBB1>S.FBFB FALSE   TRUE
7             CCX  FALSE   TRUE
8              BX   TRUE  FALSE
9           CBBFX  FALSE   TRUE
10           BFCX   TRUE  FALSE
```

Writing regular expressions for every pitch count is a tedious task and we will defer the reader to Appendix A for the full code. For the purpose of this chapter a play-by-play file containing additional information on ball/strike counts is provided.

---

[6]Applying the `nchar` function to the newly create variable `pseq` gives the number of pitches delivered in each at-bat.

7.2.4 Expected run value by count

The file `pbp11rc.csv` contains an enhanced version of the play-by-play data for the 2011 season. Other than the typical information provided by Retrosheet, this data file reports the run value for each play as calculated in Chapter 5 and additional variables such as `c00` and `c10` indicating for each possible ball/strike count whether the at-bat has gone through that particular count. This data is stored into the data frame `pbp11rc`.

```
pbp11rc <- read.csv("pbp11rc.csv")
pbp11rc[1:5, c("GAME_ID", "EVENT_ID", "c00", "c10", "c20", "c11",
  "c01", "c30", "c21", "c31", "c02", "c12", "c22", "c32",
  "RUNS.VALUE")]
```

	GAME_ID	EVENT_ID	c00	c10	c20	c11	c01	c30	c21	c31	c02	c12
1	ANA201104080	2	1	0	0	0	0	0	0	0	0	0
2	ANA201104080	3	1	0	0	1	1	0	0	0	0	1
3	ANA201104080	4	1	0	0	1	1	0	1	1	0	0
4	ANA201104080	5	1	1	0	1	0	0	0	0	0	1
5	ANA201104080	6	1	0	0	1	1	0	1	0	0	0

	c22	c32	RUNS.VALUE
1	0	0	-0.1555661
2	0	0	-0.0953746
3	0	0	0.3571207
4	1	0	-0.3347205
5	1	0	-0.3919715

For example, the at-bat in the second line of the data frame started with a 0-1 count (value 1 in column `c01`), then moved to the counts 1-1 and to 1-2 and generated a run value of -0.095. The `pbp11rc` data frame has all the necessary information to calculate the run values of the various balls/strikes counts, in the same way the value of a home run and of a single were calculated in Chapter 5.

As an illustration, one can measure the importance of getting ahead on the first pitch. The mean run value is calculated for at-bats starting with a ball and for the at-bats starting with a strike.

```
ab10 <- subset(pbp11rc, c10 == 1)
ab01 <- subset(pbp11rc, c01 == 1)
c(mean(ab10$RUNS.VALUE), mean(ab01$RUNS.VALUE))
[1]  0.03969483 -0.03546708
```

The conclusion is that the difference between a first pitch strike and a first pitch ball, as estimated with data from the 2011 season, is over 0.07 runs.

The runs value can be calculated for each possible ball/strike count. First a `data.frame` named `runs.by.count` is prepared with the twelve possible counts as done in Section 7.2 and a zero value is temporarily assigned to all counts.

```
runs.by.count <- expand.grid(balls=0 : 3, strikes = 0 : 2)
runs.by.count$value <- 0
```

A function `bs.count.run.value` is created that calculates the mean of the column `RUNS.VALUE` given the number of balls and strikes as arguments. In the function, note that in the `pbp11rc` data frame, the plate appearances are limited to the ones that have gone through the particular ball-strike count.

```
bs.count.run.value <- function(b, s){
  column.name <- paste("c", b, s, sep="")
  mean(pbp11rc[pbp11rc[, column.name] == 1, "RUNS.VALUE"])
}
```

The `mapply` function is used to apply `bs.count.run.value` to every row of the `runs.by.count` data frame, creating a new variable `value`.

```
runs.by.count$value <- mapply(FUN=bs.count.run.value,
    b=runs.by.count$balls,
    s=runs.by.count$strikes
)
```

Finally, using the `countmap` function introduced in Section 7.2 with the `runs.by.count` data frame, the run values are visualized for all of the possible balls/strikes counts. (See Figure 7.2.)

```
countmap(runs.by.count)
```

By glancing at the values and shading colors in Figure 7.2, one can construct reasonable definitions for the terms "hitter's count" and "pitcher's count." Ball/strike counts can be roughly divided in the following four categories[7]:

- Pitcher's counts: 0-2, 1-2, 2-2, 0-1;

- Neutral counts: 0-0, 1-1;

- Modest hitter's counts: 3-2, 2-1, 1-0;

- Hitter's counts: 3-0, 3-1, 2-0.

7.2.5   The importance of the previous count

In the previous section run values were calculated for any ball/strike count. In performing this calculation we simply looked at whether a plate appearance went through a particular count, without considering how it got there. In other words, we considered, for example, all the at-bats going through a 2-2

---

[7]The proposed categorization, based on an observation of Figure 7.2 reflects the one proposed by analyst Tom Tango (see `www.insidethebook.com/ee/index.php/site/comments/plate_counts/`).

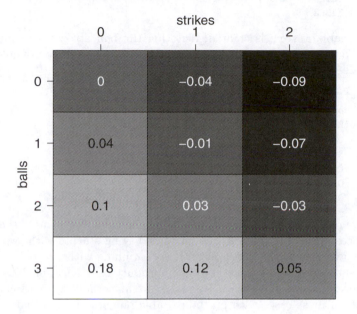

FIGURE 7.2
Run value for plate appearances through each balls/strikes count. Values estimated on data from the 2011 season.

count as having the same run expectancy, no matter if the pitcher started ahead 0-2 or fell behind 2-0. The implicit assumption in these calculations is that the previous counts have no influence on the outcome on a particular count. However, a pitcher getting ahead 0-2 is likely to "waste some pitches." That is, he would likely throw a few balls out of the strike zone with the sole intent of making the batter (who cannot afford another strike) swing at them and possibly miss or make poor contact. On the other hand, with a plate appearance starting with two balls, the batter has the luxury of not swinging at strikes in undesirable locations and wait for the pitcher to deliver a pitch of his liking.

Given the above discussion, it would seem that the run expectancy on a 2-2 count would be higher if the plate appearance started with two balls than if the pitcher started quickly with a 0-2 count. Let's investigate if there is numerical evidence to actually reflect this conjecture.

We begin by taking the subset of plays from the 2011 season that went through a 2-2 count and calculate their mean run value.

```
count22 <- subset(pbp11rc, c22 == 1)
```

```
mean(count22$RUNS.VALUE)
[1] -0.03134373
```

A new variable `after2` is created, denoting the ball/strike count after two pitches. The mean run value is calculated for each of the three possible levels of `after2`.

```
count22$after2 <- ifelse(count22$c20 == 1, "2-0",
  ifelse(count22$c02 == 1, "0-2", "1-1"))
aggregate(RUNS.VALUE ~ after2, data=count22, FUN=mean)
```

```
  after2  RUNS.VALUE
1    0-2 -0.02440420
2    1-1 -0.03277434
3    2-0 -0.03570539
```

The above results appear counterintuitive, as they seemingly imply that plate appearances going through a 2-2 count after having started with two strikes are more favorable to the hitter than those beginning with two balls.

This surprising result is actually a byproduct of a selection bias. Many plate appearances starting with two strikes end without ever reaching the 2-2 count, in most cases with an unfavorable outcome for the batter.[8] The plate appearances that survive a 0-2 count reaching 2-2 are hardly a random sample of all the plate appearances. Likely hard-to-strike-out batters are over-represented in such sample, as well as pitchers who do not posses a quality pitch to finish off opponents.

Using a similar study, comparing the paths leading to 1-1 counts gives results more in line with common sense as this count is less susceptible to the same selection bias.

```
count11 <- subset(pbp11rc, c11 == 1)
count11$after1 <- ifelse(count11$c10 == 1, "1-0", "0-1")
aggregate(RUNS.VALUE ~ after1, data=count11, FUN=mean)
```

```
  after1   RUNS.VALUE
1    0-1 -0.013234572
2    1-0 -0.009274024
```

The numbers above suggest that after reaching a 1-1 count, the batter is expected to perform slightly better if the first pitch was a ball than if it was a strike.

---

[8] In 2011, 80% of plate appearances beginning with two strikes and not reaching the 2-2 count ended with the batter making an out.

7.3 Behaviors by Count

In this section we explore how the roles of three individuals in the pitcher-batter duel are affected by the ball/strike count. How does a batter alter his swing according to being ahead or behind in the count? How does the pitcher vary the mixing of pitches according to the count? Does an umpire, more or less consciously, shrink or expand his strike zone depending on the pitch count?

An R Workspace (a file with extension .Rdata) is available containing all the datasets used in this section. Once the workspace is loaded into R, the available data frames `cabrera`, `umpires`, and `verlander` are displayed by use of the `ls` function.

```
load("balls_strikes_count.Rdata")
ls()
[1] "cabrera"   "umpires"   "verlander"
```

These datasets contain pitch-by-pitch data, including the location of pitches as recorded by Sportvision's PITCHf/x system. The `cabrera` data frame contains four years of batting data for 2012 American League Triple Crown winner Miguel Cabrera. The data frame `umpires` has information about every pitch thrown in 2012 where the home plate umpire had to judge whether it crossed the strike zone. The `verlander` data frame has four years of pitching data for 2011 Cy Young Award and MVP recipient Justin Verlander.

7.3.1 Swinging tendencies by count

It was seen in Section 7.2 that batters perform worse when falling behind in the count. For example, when there are two strikes in the count, the batter may be forced to swing at pitches he would normally let pass by to avoid being called out on strikes. Using PITCHf/x data, we explore how a very good batter like Miguel Cabrera alters his swinging tendencies according to the ball/strike count.

7.3.1.1 Propensity to swing by location

The contents of the `cabrera` data frame was introduced to the reader in Chapter 6. In this section we focus on the relationships between the variables `balls` and `strikes` indicating the count on the batter, the variables `px` and `pz` identifying the pitch location as it crosses the front of the plate, and the `swung` binary variable, denoting whether or not the batter attempted a swing on the pitch.

Using the `xyplot` function in the `lattice` package, Miguel Cabrera's swinging tendency is displayed by a scatterplot in Figure 7.3.[9].

---

[9]The provided code produces a color figure in which swings and takes are better dis-

```
sampCabrera <- cabrera[sample(1:nrow(cabrera), 500),]
topKzone <- 3.5
botKzone <- 1.6
inKzone <- -.95
outKzone <- 0.95
library(lattice)
xyplot(pz ~ px, data=sampCabrera, groups=swung,
       aspect="iso",
       xlab="horizontal location (ft.)",
       ylab="vertical location (ft.)",
       auto.key=list(points=TRUE, text=c("not swung", "swung")
                     , space="right"),
       panel=function(...){
         panel.xyplot(...)
         panel.rect(inKzone, botKzone, outKzone, topKzone,
                    border="black")
       })
```

The overlapping in the scatterplot is reduced by use of a random sample of 500 pitches from the data frame `cabrera`. From Figure 7.3, one can see that Cabrera is less likely to swing at pitches delivered farther away from the strike zone (the black box). However, it is difficult to determine Cabrera's preferred pitch location from this figure.

A contour plot is an effective alternative method to visualize batters' swinging preferences. The plot is used to visualize three dimensional data on a two-dimensional surface. Widely used in cartography and meteorology, the contour plot usually features spatial coordinates as the first two variables, while the third variable (which can be, for example, elevation in cartography or barometric pressure in meteorology) is plotted as a contour line, also called an isopleth. The contour line is a curve joining points sharing equal values of the third variable.

As a first step in producing a contour plot, a polynomial surface is fit using the `loess` function using the horizontal and vertical locations `px` and `pz` as predictors and `swung` as the outcome variable. The output of this fit is stored in the variable `miggy.loess`.

```
miggy.loess <- loess(swung ~ px + pz, data=cabrera,
  control=loess.control(surface="direct"))
```

After the surface on the dataset has been fit, we are interested in predicting the likelihood of a swing by Cabrera at various pitch locations. Using the `expand.grid` function, a data frame is built consisting of combinations of horizontal locations from -2 (two feet to the left of the middle of home plate) to $+2$ (two feet to the right of the middle of the plate) and vertical locations

---

cerned than in Figure 7.3 If one wants to obtain the black and white version displayed in this book, the following line of code has to be inserted before the calling of `xyplot`: `trellis.par.set(canonical.theme(color=FALSE))`

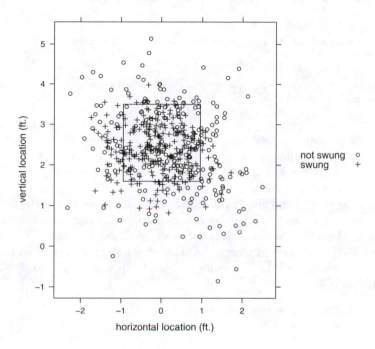

FIGURE 7.3

Scatterplot of Miguel Cabrera's swinging tendency by location. Sample of 500 pitches. View from the catcher's perspective.

from the ground (value of zero) to six feet of height, using subintervals of 0.1 feet. By using the `predict` function,[10] the likelihood of Miguel's swing is obtained at every location in the data frame.

```
pred.area <- expand.grid(px=seq(-2, 2, 0.1), pz=seq(0, 6, 0.1))
pred.area$fit <- c(predict(miggy.loess, pred.area))
```

From the data frame `pred.area` the likelihood that Miguel will swing is estimated for three different locations – a pitch down the middle and two and a half feet from the ground ("down Broadway"), a ball that hits the ground in the middle of the plate ("ball in the dirt"), and another one delivered at midheight (2.5 feet from the ground) but way outside (two feet from the middle of the plate). In each case, the `subset` function is used to take a subset of the prediction data frame `pred.area` with specific values of the horizontal and vertical locations `px` and `pz`.

---

[10]The c function used in the second assignment converts the `matrix` resulting from the `predict` call into a vector.

```
subset(pred.area, px == 0 & pz == 2.5)   #down Broadway
      px  pz        fit
1046   0 2.5 0.8202528
subset(pred.area, px == 0 & pz == 0)   #ball in the dirt
   px pz        fit
21  0   0 0.1729128
subset(pred.area, px == 2 & pz == 2.5)   #way outside
      px  pz        fit
1066   2 2.5 0.06238047
```

The results are quite consistent with what one would expect: the pitch right in the heart of the strike zone generates Cabrera's swing more than 80 percent of the time, while the ball in the dirt and the ball outside generates a swing at 17 percent and six percent rates, respectively.

A contour plot of the likelihood of the swing as a function of the horizontal and vertical locations of the pitch is constructed using the `contourplot` function in the `lattice` package. The meaning of most arguments in the R code has been explained in Chapter 6; the `at` parameter is used to indicate the levels at which we want the contour lines to be drawn. Figure 7.4 shows the resulting contour plot.

```
contourplot(fit ~ px * pz, data=pred.area,
            at=c(.2, .4, .6, .8),
            aspect="iso",
            xlim=c(-2, 2), ylim=c(0, 5),
            xlab="horizontal location (ft.)",
            ylab="vertical location (ft.)",
            panel=function(...){
              panel.contourplot(...)
              panel.rect(inKzone, botKzone, outKzone, topKzone,
                      border="black", lty="dotted")
            })
```

As expected, the likelihood of a swing decreases the further the ball is delivered from the middle of the strike zone. The plot also shows that Cabrera has a tendency to swing at pitches on the inside part of the plate.

7.3.1.2   Effect of the ball/strike count

Figure 7.4 reports Miguel's swinging tendency over all pitch counts. Can we visualize how Cabrera varies his approach according to the ball/strike count? Specifically, does Cabrera become more selective when he is ahead and can afford to wait for a pitch of his liking and, conversely, does he "expand his zone" when there are two strikes and he cannot allow to let another called strike go by? The process of calculating the swing propensity by location has been described in Section 7.3.1.1. Here the same process is performed on a subset of data, specifically on pitches delivered on a 0-0 count.

In the following R code, a new variable `bscount` is defined giving the

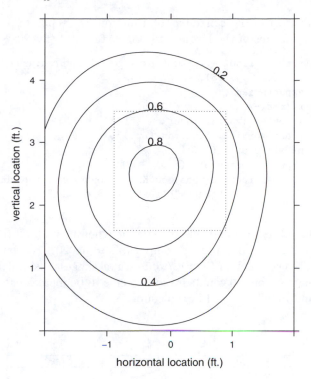

FIGURE 7.4
Contour plot of Miguel Cabrera's swinging tendency by location, where the
view from the catcher's perspective. The contour lines are labeled by the
probability of swinging at the pitch.

ball-strike count. Using the `subset` function, a new data frame `miggy00` is
constructed consisting of pitch data only when the ball-strike out is 0-0. The
`loess` function is used to find the likelihood of swing surface, and the `predict`
function computes the likelihood for every horizontal and vertical location in
the data frame `pred.area`.

```
cabrera$bscount <- paste(cabrera$balls, cabrera$strikes, sep="-")
miggy00 <- subset(cabrera, bscount == "0-0")
miggy00loess <- loess(swung ~ px + pz, data=miggy00, control=
  loess.control(surface="direct"))
pred.area$fit00 <- c(predict(miggy00loess, pred.area))
```

The same procedure was repeated for the 0-2 and 2-0 counts. The code
is not shown here, but the reader can write the R code by slight modifications of the code for the 0-0 count. Once the additional variables `fit02` and
`fit20` were obtained for the `pred.area` data frame, contour plots on separate
panels can be constructed to compare Cabrera's swinging tendencies by pitch

count (Figure 7.5.) In the `contourplot` function, the separate panels display is produced by means of the formula `fit00 + fit02 + fit20 ∼ px * pz`.

```
contourplot(fit00 + fit02 + fit20 ~ px * pz, data=pred.area,
            at=c(.2, .4, .6),
            aspect="iso",
            xlim=c(-2, 2), ylim=c(0, 5),
            xlab="horizontal location (ft.)",
            ylab="vertical location (ft.)",
            panel=function(...){
              panel.contourplot(...)
              panel.rect(inKzone, botKzone, outKzone, topKzone,
                         border="black", lty="dotted")
})
```

As expected, Cabrera expands his swing zone when behind 0-2 (his 40% contour line on 0-2 counts has an area comparable to his 20% contour line on 0-0 counts). The third panel on Figure 7.5 does not suggest a shrinkage of Cabrera's swinging zone on 2-0 counts. Miguel seems to be increasingly looking for pitches up-and-in when ahead in the count.

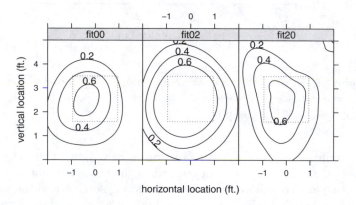

FIGURE 7.5
Miguel Cabrera's 50/50 swing zone in different balls/strikes counts. View from the catcher's perspective.

7.3.2   Pitch selection by count

We now move to the other side of the pitcher/batter duel in our learning about the effect of the pitch count. Pitchers generally possess arsenals of two to five different pitch types. Nearly all pitchers have a fastball at their disposal, which is generally a pitch that is easy to locate at the desired location. So-called complimentary pitches, such as curve balls or sliders, while often effective (especially when hitters are not expecting them), are harder to control and rarely used by pitchers behind in the count. In this section we look at one pitcher (arguably one of the best in MLB at the time of this writing) and explore how he chooses from his pitch repertoire according to the ball/strike count.

The `verlander` data frame, consisting of over 15 thousand observations, consists of pitch data for Justin Verlander for five seasons. Using the `table` command, a frequency table of the types of pitches Verlander has thrown since 2009 is obtained.

```
table(verlander$pitch_type)
```

```
  CH   CU   FF   FT   SL
2550 2716 6756 2021 1264
```

As is the case with most major league pitchers, Verlander most frequently uses the fastball. He uses two variations of a fastball, a four-seamer (FF) and a two-seamer (FT). He complements his fastballs with a curve ball (CU) ,a change-up (CH), and a slider (SL).

The `prop.table` function is used to obtain the pitch type proportions rather than their frequencies. The input to `prop.table` is the table of frequencies. To obtain percentages, the result is multiplied by 100 and rounded (using the `round` function) to the nearest integer.

```
round(100 * prop.table(table(verlander$pitch_type)))
```

```
CH CU FF FT SL
17 18 44 13  8
```

It can be seen from the table that 44% of Verlander's pitches during this five-season period were four-seamers.

Before moving to exploring pitch selection by ball/strike count, a frequency table is used to explore the pitch selection by batter handedness. One constructs frequencies of pitch type for each batter hand by specifying two parameters in the `table` function. To compute proportions of pitch type for each batter hand, the `prob.table` function is used with argument `margin=2`. (The function computes proportions by row if `margin=1` and by column if `margin=2`.)

```
type_verlander_hand <- with(verlander, table(pitch_type,
  batter_hand))
round(100 * prop.table(type_verlander_hand, margin=2))
```

```
          batter_hand
pitch_type  L   R
        CH 23   8
        CU 17  18
        FF 43  45
        FT 15  11
        SL  2  17
```

Note that the pitch selection is quite different depending on the handedness of the opposing batter. In particular, the right-handed Verlander uses his change-up nearly a quarter of the time against left-handed hitters, but only eight percent of the time against right-handed hitters. Conversely the slider is nearly absent from his repertoire when he faces lefties, while he uses it close to one out of five times against righties.

Batter-hand differences in pitch selection are common among major league pitchers and they exist because the effectiveness of a given pitch depends on the handedness of the pitcher and the batter. The slider and change-up comparison is a typical example, a slider is very effective against batters of the same handedness and a change-up can be successful when facing opposite-handed batters.

Justin's pitch selection can be explored by pitch count. First, a new `bscount` variable is created by merging the values in the columns `balls` and `strikes`. Since pitch selection depends on batter handedness, pitch mixing is examined separately by the opponent handedness. In the following code, the `subset` function is used to select Verlander's pitches delivered to right-handed batters. The `table` function constructs a table of frequencies by count and pitch type and the `prop.table` function with the `margin=1` argument computes row proportions.

```
verlander$bscount <- paste(verlander$balls, verlander$strikes,
  sep="-")
verl_RHB <- subset(verlander, batter_hand == "R")
verl_type_cnt_R <- table(verl_RHB$bscount, verl_RHB$pitch_type)
round(100 * prop.table(verl_type_cnt_R, margin=1))
```

```
      CH CU FF FT SL
0-0    7 11 53 16 13
0-1    6 24 40 10 19
0-2   16 28 27  6 22
1-0    5 11 52 11 21
1-1    8 24 40 10 19
1-2   14 33 28  6 19
2-0    2  2 70 15 12
2-1    6  8 51 14 20
2-2   10 29 36  9 16
3-0    8  0 81 10  0
3-1    2  0 78 12  8
3-2    4  4 69 12 11
```

The effect of the ball/strike count on the choice of pitches is apparent when comparing pitcher's counts and hitter's counts. When behind 2-0, Verlander uses his four-seamer seven times out of ten; the percentage goes up to 78% when trailing 3-1 and 81% on 3-0 counts. Conversely, when Justin has the chance to strike the batter out, the use of the four-seamer diminishes. In fact he throws it less than 30 percent of the time both on 0-2 and 1-2 counts. On a full count, Verlander's propensity to go with the fastball is similar to the one in hitters' counts – this is consistent with the numbers in Figure 7.2 that indicate the 3-2 count being slightly favorable to the hitter. Similarly, one can explore Verlander's choices by count when facing a left-handed hitter.

7.3.3   Umpires' behavior by count

Hardball Times author John Walsh wrote a 2010 article *The Compassionate Umpire,*[11] in which he showed that home plate umpires tend to modify their ball/strike calling behavior by slightly favoring the player who is behind in the count. In other words, umpires tend to enlarge their strike zone in hitter's counts and to shrink it when pitchers are ahead. In this section we visually explore John's finding, by plotting contour lines for three different ball/strike counts.

The `umpires` data frames are similar to the `verlander` and `cabrera` ones. A sample of its contents, obtained by typing the command `umpires[sample(1:nrow(umpires), 20),]` in the R console is shown in Table 7.2. The data consist of every pitch of the 2012 season for which the home plate umpire had to judge whether it crossed the strike zone. Additional columns not present in either the `verlander` or the `cabrera` data frames identify the name of the umpire (variable `umpire`) and whether the pitch was called for a strike (variable `called_strike`).

We proceed similarly to the analysis of Section 7.3.1.2, using the `loess` function to estimate the umpires' likelihood of calling a strike, based on the location of the pitch. Following is the R code for the 0-0 count. Note that the analysis is limited to plate appearances featuring right-handed batters, as it has been shown that umpires tend to call pitches slightly differently depending on the handedness of the batter. In addition, the `loess` smoother is applied on a subset of 3000 randomly selected pitches, to facilitate the computation time of the script.

```
umpiresRHB <- subset(umpires, batter_hand == "R")
ump00 <- subset(umpiresRHB, balls == 0 & strikes == 0)
ump00smp <- ump00[sample(1:nrow(ump00), 3000),]
ump00.loess <- loess(called_strike ~ px + pz, data=ump00smp,
  control=loess.control(surface="direct"))
```

By slightly modifying the code above, the reader can easily repeat the process for other counts. In this section we compare the 0-0 count to the most extreme batter and pitcher counts, 3-0 and 0-2 counts, respectively.

In this instance the contour lines of balls/strikes calling in different counts will be plotted in a single panel, rather than in separate panels as was done for swinging tendencies in Figure 7.5. The `contourLines` function[12] is helpful for this task, as it calculates coordinates to plot contour lines. The following code calculates, for the 0-0 count, the 50% strike calling contour line and

---

[11] www.hardballtimes.com/main/article/the-compassionate-umpire/.

[12] The function comes with the `grDevices` package, which is loaded by default in R.

TABLE 7.2

A twenty rows sample of the umpires dataset.

(row #)	season	umpire	batter_hand	pitch_type	balls	strikes	px	pz	called_strike
80357	2012	Doug Eddings	R	SL	0	0	-0.17	2.88	1
33199	2012	Alfonso Marquez	L	FF	0	0	0.03	2.53	1
187437	2012	Angel Hernandez	L	SL	1	1	-1.48	2.35	0
260385	2012	Ed Rapuano	R	FT	1	1	-1.44	2.05	0
169709	2012	Sam Holbrook	L	FF	2	1	0.80	2.42	0
195437	2012	Dan Bellino	L	FT	0	0	1.07	1.59	0
166333	2012	Eric Cooper	R	FF	0	0	-0.77	2.35	1
264734	2012	Gerry Davis	L	FF	0	0	0.55	2.08	1
191453	2012	D.J. Reyburn	R	FC	1	1	-1.13	2.69	1
10811	2012	Kerwin Danley	R	SL	1	1	0.89	2.89	0
12474	2012	Mark Ripperger	R	FF	0	0	-1.11	1.80	0
134379	2012	Laz Diaz	L	FF	0	0	-1.84	3.21	0
55770	2012	Jeff Nelson	R	CU	2	2	1.70	2.02	0
203183	2012	Tim Timmons	R	FF	1	0	1.18	2.97	0
202977	2012	Sam Holbrook	L	SL	2	1	-0.23	1.01	0
58915	2012	Mark Wegner	L	CH	0	1	-1.65	2.42	0
300411	2012	Tim Timmons	L	FF	0	0	-1.21	2.40	0
84853	2012	Paul Nauert	R	CU	3	2	-0.23	1.04	0
312177	2012	Jim Joyce	L	SL	3	2	-0.25	1.65	0
216589	2012	Brian Runge	L	FT	1	1	-0.66	2.29	1

converts the resulting list into a data frame, to which a new variable `bscount` indicating the balls/strikes count is added.

```
ump00contour <- contourLines(x=seq(-2, 2, 0.1),
                             y=seq(0, 6, 0.1),
                             z=predict(ump00.loess, pred.area),
                             levels=c(.5))
ump00df <- as.data.frame(ump00contour)
ump00df$bscount <- "0-0"
```

The function `contourLines` from the `grDevices` package (which is loaded by default in R) requires two vectors (`x` and `y`) of values in ascending order, that combined define the locations where the values are measured. The values are stored in a matrix `z` whose dimensions are given by `length(x)` × `length(y)`. Then either a number of levels (`nlevels`) can be specified, or a vector (`levels`) which indicates the levels at which to draw contour lines. The function returns values that can then be supplied to plotting functions in order to draw the contour lines.

The 50% strike calling contour lines for the counts of interest can be plotted as line plots. The output produced by the following code (after `ump02df` and `ump30df` are created similarly to `ump00df`) can be seen in Figure 7.6.

```
umpireContours <- rbind(ump00df, ump02df, ump30df)
trellis.par.set(theme=canonical.theme(color=FALSE))
myKey <- list(lines=TRUE
              , points=FALSE
              , space="right"
              , title="balls/strikes count"
              , cex.title=1
              , padding=4)
xyplot(y ~ x , data=umpireContours
       , groups=bscount
       , type="l", aspect="iso"
       , col="black"
       , xlim=c(-2, 2), ylim=c(0, 5)
       , xlab="horizontal location (ft.)"
       , ylab="vertical location (ft.)"
       , auto.key=myKey
       , panel=function(...){
         panel.xyplot(...)
         panel.rect(inKzone, botKzone, outKzone, topKzone,
                    border="grey70", lwd=2)
       })
```

This figure shows that the umpire's strike zone is shrunk in a 0-2 pitch count, and slightly expanded in a 3-0 count.

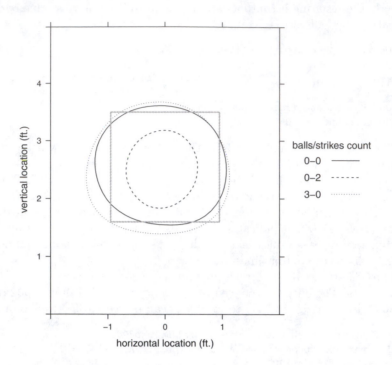

FIGURE 7.6
Umpires' 50/50 strike calling zone in different balls/strikes counts viewed from the catcher's perspective.

7.4 Further Reading

Palmer (1983) is possibly one of the first examinations of the balls/strikes count effect on the outcome of plate appearances: it is based on data from World Series games from 1974 to 1977 and features a table resembling Figures 7.1 and 7.2. Walsh (2008) calculates the run value of a ball and of a strike at every count and uses the results for ranking baseball's best fastballs, sliders, curveballs, and change-ups. Walsh (2010) shows how umpires are (perhaps unconsciously) affected by the balls/strikes count when judging pitches. In particular, he presents a scatterplot showing a very high correlation between the strike zone area and the count run value (see Figure 7.2). Allen (2009a, 2009b) and Marchi (2010) illustrate so-called platoon splits (i.e. the different effectiveness against same-handed versus opposite-handed batters) for various pitch types.

7.5 Exercises

1. (**Run Value of Individual Pitches**)

   (a) Calculate the run value of a ball and of a strike at any count. For 3-ball and 2-strike counts you need the value of a walk and a strike-out respectively (you can calculate them as done for other events in Chapter 5).

   (b) Compare your values to the ones proposed by John Walsh in the article www.hardballtimes.com/main/article/searching-for-the-games-best-pitch/.

2. (**Length of Plate Appearances**)

   (a) Calculate the length, in term of pitches, of the average plate appearance by batting position using Retrosheet data for the 2011 season.

   (b) Does the eighth batter in the National League behave differently than his counterpart in the American League?

   (c) Repeat the calculations in (a) and (b) for the 1991 and 2011 seasons and comment on any differences between the seasons that you find.

3. (**Pickoff Attempts**)

   Identify the baserunners who, in the 2011 season, drew the highest number of pickoff attempts when standing at first base with second base unoccupied.

4. (**Umpire's Strike Zone**)

   By drawing a contour plot, compare the umpire's strike zone for left-handed and right-handed batters. Use only the rows of the data frame where the pitch type is a four-seam fastball.

5. (**Umpire's Strike Zone, Continued**)

   By drawing one or more contour plots, compare the umpire's strike zone by pitch type. For example, compare the 50/50 contour lines of four-seam fastballs and curveballs when a right-handed batter is at the plate.

8

Career Trajectories

CONTENTS

8.1	Introduction	....	187
8.2	Mickey Mantle's Batting Trajectory	....	188
8.3	Comparing Trajectories	....	192
	8.3.1 Some preliminary work	....	192
	8.3.2 Computing career statistics	....	194
	8.3.3 Computing similarity scores	....	195
	8.3.4 Defining age, OBP, SLG, and OPS variables	....	197
	8.3.5 Fitting and plotting trajectories	....	198
8.4	General Patterns of Peak Ages	....	202
	8.4.1 Computing all fitted trajectories	....	202
	8.4.2 Patterns of peak age over time	....	203
	8.4.3 Peak age and career at-bats	....	204
8.5	Trajectories and Fielding Position	....	205
8.6	Further Reading	....	208
8.7	Exercises	....	209

8.1 Introduction

The R system is well-suited for fitting statistical models to data. One popular topic in sabermetrics is the rise and fall of a player's season batting, fielding, or pitching statistics from his MLB debut to retirement. Generally, it is believed that most players peak in their late 20s, although some players tend to peak at later ages. A simple way of modeling a player's trajectory is by means of a quadratic or parabolic curve. By use of the `lm` (linear model) function in R, it is straightforward to fit this model using the player's age and his OPS statistics.

We begin in Section 8.2 by considering a famous career trajectory. Mickey Mantle made an immediate impact on the New York Yankees at age 19 and quickly matured into one of the best hitters in baseball. But injuries took a toll on Mantle's performance and his hitting declined until his retirement at age 36. We use Mantle to introduce the quadratic model – using this model, one can define his peak age, the maximum performance, and the rate of increase and decline in performance.

To compare career performances of similar players, it is helpful to contrast their trajectories and Section 8.3 illustrates the computation of many fitted trajectories. Using Bill James' notion of similarity scores, we write a function that will find players who are most similar to a given hitter. Then we graphically compare the OPS trajectories of these similar players; by viewing these graphs we gain a general understanding of the possible trajectory shapes.

A general problem focuses on a player's peak age. In Section 8.4, we look at the fitted trajectories of all hitters with at least 2000 career at-bats. The pattern of peak ages across eras and as a function of the number of career at-bats is explored. Also, since it is common to compare players who play the same position, in Section 8.5 we focus on the period 1985-1995 and contrast the peak ages for players who play different fielding positions.

8.2   Mickey Mantle's Batting Trajectory

To start looking at career trajectories, we consider batting data from the great slugger Mickey Mantle. To obtain his season-by-season hitting statistics, the Lahman data files `Master.csv` and `Batting.csv` are read into R and saved in the data frames `Master` and `Batting`.

```
Batting <- read.csv("Batting.csv")
Master <- read.csv("Master.csv")
```

We extract Mantle's `playerID` from the `Master` data frame. By use of the `subset` function, the line in the `Master` data file is found where `nameFirst ==` `"Mickey` and `nameLast == "Mantle"`. His player id is stored in the variable `mantle.id`.

```
mantle.info <- subset(Master,
  nameFirst == "Mickey" & nameLast == "Mantle")
mantle.id <- as.character(mantle.info$playerID)
```

One small complication is that certain statistics such as SF and HBP were not recorded for older seasons and are currently coded as `NA`. A convenient way of recoding these missing values to 0 is by the `recode` function in the `car` package.

```
library(car)
Batting$SF <- recode(Batting$SF, "NA = 0")
Batting$HBP <- recode(Batting$HBP, "NA = 0")
```

To compute Mantle's age for each season, we need to know his birth year which is available in the `Master` data frame. Major League Baseball defines a player's age as his age on June 30 of that particular season. To facilitate the computation of ages, a new function `get.birthyear` is defined which gives the "official" birth year of a player with id `player.id`, similar to what was done in the `getinfo` function of Section 3.8.

```
get.birthyear <- function(player.id){
  playerline <- subset(Master, playerID == player.id)
  birthyear <- playerline$birthYear
  birthmonth <- playerline$birthMonth
  ifelse(birthmonth >= 7, birthyear + 1, birthyear)
}
```

To check this function, the MLB birth year for Mantle is found using his id stored in `mantle.id`.

```
get.birthyear(mantle.id)
[1] 1932
```

Mantle's batting statistics are obtained by means of the user-defined function `get.stats`. The inputs are the player id and the output is a data frame containing the player's hitting statistics. This function computes the player's age (variable `Age`) for all seasons, and also computes the player slugging percentage (`SLG`), on-base percentage (`OBP`), and OPS for all seasons. Note that the function `get.birthyear` is used to find the player's MLB birth year.

```
get.stats <- function(player.id){
  d <- subset(Batting, playerID==player.id)
  byear <- get.birthyear(player.id)
  d$Age <- d$yearID - byear
  d$SLG <- with(d, (H - X2B - X3B - HR +
                      2 * X2B + 3 * X3B + 4 * HR) / AB)
  d$OBP <- with(d, (H + BB) / (H + AB + BB + SF))
  d$OPS <- with(d, SLG + OBP)
  d
}
```

After reading the function `get.stats` into R, we obtain Mantle's statistics by applying this function with input `mantle.id` – the resulting data frame of hitting statistics is stored in `Mantle`.

```
Mantle <- get.stats(mantle.id)
```

A good measure of batting performance is OPS, the sum of a player's slugging percentage and his on-base percentage. How does Mantle's OPS season values vary as a function of his age? To address this question, the `plot` function is used to construct a scatterplot of OPS against age. (See Figure 8.1.)

```
with(Mantle, plot(Age, OPS, cex=1.5, pch=19))
```

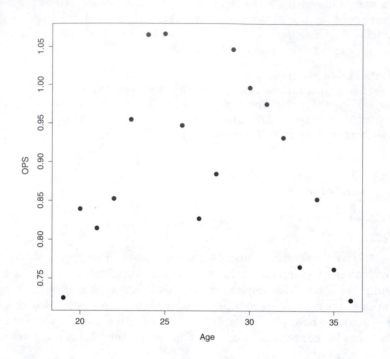

FIGURE 8.1
Scatterplot of OPS against age for Mickey Mantle.

Looking at this figure, it is clear that Mantle's OPS values tend to increase from age 19 to his late 20s, and then generally decrease until his retirement at age 36. One can model this up-and-down relationship by use of a smooth curve. This curve will help us understand and summarize Mantle's career batting trajectory and will make it easier to compare Mantle's trajectory with other players with similar batting performances.

A convenient choice of smooth curve is a quadratic function of the form

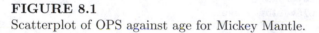

$$A + B(Age - 30) + C(Age - 30)^2,$$

where the constants A, B, and C are chosen so that curve is the "best" match to the points in the scatterplot. This quadratic curve has the following nice properties that make it easy to use.

1. The constant A is the predicted value of OPS when the player is 30 years old.

2. The function reaches its largest value at

$$PEAK.AGE = 30 - \frac{B}{2C}.$$

This is the age where the player is estimated to have his peak batting performance during his career.

3. The maximum value of the curve is

$$MAX = A - \frac{B^2}{4C}.$$

This is the estimated largest OPS of the player over his career.

4. The coefficient C, typically a negative value, tells us about the degree of curvature in the quadratic function. If a player has a "large" value of C, this indicates that he more rapidly reaches his peak level and more rapidly decreases in ability until retirement. One simple interpretation is that C represents the change in OPS from his peak age to one year later.

A new function `fit.model` is written to fit this quadratic curve to a player's batting data. The input to this function is a data frame `d` containing the player's batting statistics including the variables `Age` and `OPS`. The function `lm` is used to fit the quadratic curve – the formula

$$\text{OPS} \sim \text{I(Age - 30)} + \text{I((Age - 30)}^2)$$

indicates that `OPS` is the response and (`Age` - 30) and (`Age` - 30)^2 are the predictors. The estimated coefficients A, B, and C are saved in the vector `b`. The peak age and maximum value are stored in the variables `Age.max` and `Max`.

```
fit.model <- function(d){
  fit <- lm(OPS ~ I(Age - 30) + I((Age - 30)^2), data=d)
  b <- coef(fit)
  Age.max <- 30 - b[2] / b[3] / 2
  Max <- b[1] - b[2] ^ 2 / b[3] / 4
  list(fit=fit,
       Age.max=Age.max, Max=Max)
}
```

The function `fit.model` is applied to Mantle's data frame – the output of this function includes the object `fit` that stores all of the calculations of the quadratic fit. In addition, this function outputs the peak age and maximum value displayed in the following code.

```
F2 <- fit.model(Mantle)
coef(F2$fit)
    (Intercept)      I(Age - 30) I((Age - 30)^2)
     0.955433417     -0.020289562    -0.003520738
c(F2$Age.max, F2$Max)
I(Age - 30) (Intercept)
   27.118564    0.984665
```

The best fitting curve is given by

$$0.955 - 0.0202(Age - 30) - 0.00352(Age - 30)^2,$$

Using this model, Mantle peaked at age 27 and his maximum OPS for the curve is estimated to be 0.985. The estimated value of the curvature parameter is -0.00352, thus Mantle's decrease in OPS between his peak age and one year older is 0.00352.

This best quadratic curve is placed on the scatterplot. The `predict` function is used to estimate Mantle's OPS from the curve for the sequence of age values and the `lines` function overlays these values as a line on the current plot. Two applications of `abline` show the locations of the peak age and the maximum, and the `text` function is used to label these values. The resulting graph is displayed in Figure 8.2.

```
lines(Mantle$Age, predict(F2$fit, Age=Mantle$Age), lwd=3)
abline(v=F2$Age.max, lwd=3, lty=2, col="grey")
abline(h=F2$Max, lwd=3, lty=2, col="grey")
text(29, .72, "Peak.age" , cex=2)
text(20, 1, "Max", cex=2)
```

Although the focus was on the best fitting quadratic curve, more details about the fitting procedure are stored in the output of `lm` that is stored in the variable `fit`. We display part of the output display by finding the `summary` of the fit.

```
summary(F2$fit)
...
Residual standard error: 0.07501 on 15 degrees of freedom
Multiple R-squared: 0.6093,        Adjusted R-squared: 0.5572
F-statistic: 11.69 on 2 and 15 DF,  p-value: 0.0008692
```

The value of R^2 is 0.6093 – this means that approximately 61% of the variability in Mantle's OPS values can be explained by the quadratic curve. The residual standard error is equal to 0.075. Approximately 2/3 of the vertical deviations (the "residuals") from the curve fall between plus and minus one residual standard error. In this case, the interpretation is that approximately 2/3 of the residuals fall between -0.075 and 0.075.

8.3   Comparing Trajectories

8.3.1   Some preliminary work

When we think about hitting trajectories of players, one relevant variable seems to be a player's fielding position. Hitting expectations of a catcher, an

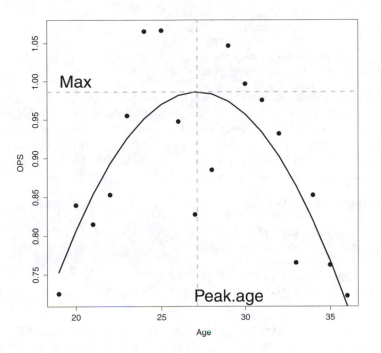

FIGURE 8.2
Scatterplot of OPS against age for Mickey Mantle with a quadratic fit added.
The location of the peak age and the maximum OPS fit are displayed.

important defensive position, may be different from the hitting expectations
of a first baseman. To compare trajectories of players with the same position,
fielding position should be recorded in our database.

The data frame `Batting` has already been created. We read in the fielding
data file `Fielding.csv` and store it in the data frame `Fielding`.

```
Fielding <- read.csv("Fielding.csv")
```

Many players in the history of baseball have had short careers and in our
study of trajectories, it seems reasonable to limit our analysis to players who
have had a minimum of at-bats. We consider only players with 2000 at-bats –
this will remove hitting data of pitchers and other players with short careers.
To take this subset of the `Batting` data frame, we use the `ddply` function
in the `plyr` package to compute the career at-bats for all players – the new
variable is called `Career.AB`. By use of the `merge` function, we add this new
variable to the `Batting` data frame. Finally, using the `subset` function, a new
data frame `Batting.2000` is created consisting only of the "minimum 2000
AB" hitters.

```
library(plyr)
AB.totals <- ddply(Batting, .(playerID),
                   summarize,
                   Career.AB=sum(AB, na.rm=TRUE))
Batting <- merge(Batting, AB.totals)
Batting.2000 <- subset(Batting, Career.AB >= 2000)
```

To add fielding information to our data frame, a new function
`find.position` is written that finds the primary fielding position for a given
player. In the function, the `subset` function is used to extract the lines of the
`Fielding` data frame for a player with player id equal to `p`. The number of
games played at each possible position are tallied, and the function returns
the position where the most games are played.[1]

```
find.position <- function(p){
  positions <- c("OF", "1B", "2B", "SS", "3B", "C", "P", "DH")
  d <- subset(Fielding, playerID == p)
  count.games <- function(po)
    sum(subset(d, POS == po)$G)
  FLD <- sapply(positions, count.games)
  positions[FLD == max(FLD)][1]
}
```

In the following code, `PLAYER` is a character vector of player ids for all players
with at least 2000 career at-bats. The `sapply` function is used to find the primary fielding position for all players and the new data frame `Fielding.2000`
is created using the function `data.frame` containing the player ids and the
fielding positions (variable `POS`). This new information is placed into the
`Batting.2000` data frame by use of the `merge` function.

```
PLAYER <- as.character(unique(Batting.2000$playerID))
POSITIONS <- sapply(PLAYER, find.position)
Fielding.2000 <- data.frame(playerID=names(POSITIONS),
                            POS=POSITIONS)
Batting.2000 <- merge(Batting.2000, Fielding.2000)
```

8.3.2   Computing career statistics

Groups of similar hitters will be found on the basis of their career statistics.
Towards this goal, one needs to compute the career games played, at-bats,
runs, hit, etc., for each player in the `Batting.2000` data frame. This is conveniently done using the `ddply` function. In the R code, the `summarize` argument
option is used to find the sum of different batting statistics for each hitter.
A new data frame `C.totals` is created with the player id variable `playerID`
and new career variables `C.G`, `C.AB`, `C.R`, and so on.

---

[1]In the rare case where there are two or more positions with the most games played, the
function will take the first position.

```
library(plyr)
C.totals <- ddply(Batting.2000, .(playerID),
                  summarize,
                  C.G=sum(G, na.rm=TRUE),
                  C.AB=sum(AB, na.rm=TRUE),
                  C.R=sum(R, na.rm=TRUE),
                  C.H=sum(H, na.rm=TRUE),
                  C.2B=sum(X2B, na.rm=TRUE),
                  C.3B=sum(X3B, na.rm=TRUE),
                  C.HR=sum(HR, na.rm=TRUE),
                  C.RBI=sum(RBI, na.rm=TRUE),
                  C.BB=sum(BB, na.rm=TRUE),
                  C.SO=sum(SO, na.rm=TRUE),
                  C.SB=sum(SB, na.rm=TRUE))
```

In the new data frame, we compute each player's career batting average `C.AVG` and his career slugging percentage `C.SLG`.

```
C.totals$C.AVG <- with(C.totals, C.H / C.AB)
C.totals$C.SLG <- with(C.totals,
        (C.H - C.2B - C.3B - C.HR + 2 * C.2B +
          3 * C.3B + 4 * C.HR) / C.AB)
```

The career statistics data frame `C.totals` is merged with the fielding data frame `Fielding.2000`. Each fielding position has an associated value, and a series of `ifelse` functions are used to define a value position variable `Value.POS` from the position variable `POS`.

```
C.totals <- merge(C.totals, Fielding.2000)
C.totals$Value.POS <- with(C.totals,
                     ifelse(POS == "C", 240,
                     ifelse(POS == "SS", 168,
                     ifelse(POS == "2B", 132,
                     ifelse(POS == "3B", 84,
                     ifelse(POS == "OF", 48,
                     ifelse(POS == "1B", 12, 0)))))))
```

8.3.3 Computing similarity scores

Bill James introduced the concept of similarity scores to facilitate the comparison of players on the basis of career statistics. To compare two hitters, one starts at 1000 points and subtracts points based on the differences in different statistical categories. One point is subtracted for each of the following differences: (1) 20 games played, (2) 75 at-bats, (3) 10 runs scored, (4) 15 hits, (5) 5 doubles, (6) 4 triples, (7) 2 home runs, (8) 10 runs batted in, (9) 25 walks, (10) 150 strikeouts, (11) 20 stolen bases, (12) 0.001 in batting average, and (13) 0.002 in slugging percentage. In addition, one subtracts the absolute value of the difference between the fielding position values of the two players.

The function `similar` will find the players most similar to a given player using similarity scores on career statistics and fielding position. One inputs the id for the particular player and the number of similar players to be found (including the given player). The output is a data frame of player statistics, ordered in decreasing order by similarity scores.

```
similar <- function(p, number=10){
  P <- subset(C.totals, playerID == p)
  C.totals$SS <- with(C.totals,
               1000 -
               floor(abs(C.G - P$C.G) / 20) -
               floor(abs(C.AB - P$C.AB) / 75) -
               floor(abs(C.R - P$C.R) / 10) -
               floor(abs(C.H - P$C.H) / 15) -
               floor(abs(C.2B - P$C.2B) / 5) -
               floor(abs(C.3B - P$C.3B) / 4) -
               floor(abs(C.HR - P$C.HR) / 2) -
               floor(abs(C.RBI - P$C.RBI) / 10) -
               floor(abs(C.BB - P$C.BB) / 25) -
               floor(abs(C.SO - P$C.SO) / 150) -
               floor(abs(C.SB - P$C.SB) / 20) -
               floor(abs(C.AVG - P$C.AVG) / 0.001) -
               floor(abs(C.SLG - P$C.SLG) / 0.002) -
               abs(Value.POS - P$Value.POS))
  C.totals <- C.totals[order(C.totals$SS, decreasing=TRUE), ]
  C.totals[1:number, ]
}
```

To illustrate the use of this function, suppose one is interested in finding the five players who are most similar to Mickey Mantle. Recall that the player id for Mantle is stored in the variable `mantle.id`. We use the function `similar` with inputs `mantle.id` and 6.

```
similar(mantle.id, 6)
        playerID  C.G C.AB  C.R  C.H C.2B C.3B C.HR C.RBI C.BB C.SO C.SB
1282 mantlmi01 2401 8102 1677 2415  344   72  536  1509 1733 1710  153
1308 matheed01 2391 8537 1509 2315  354   72  512  1453 1444 1487   68
1805 schmimi01 2404 8352 1506 2234  408   59  548  1595 1507 1883  174
1844 sheffga01 2576 9217 1636 2689  467   27  509  1676 1475 1171  253
2013 thomafr04 2322 8199 1494 2468  495   12  521  1704 1667 1397   32
1900  sosasa01 2354 8813 1475 2408  379   45  609  1667  929 2306  234
          C.AVG      C.SLG POS Value.POS   SS
1282 0.2980745 0.5567761  OF        48 1000
1308 0.2711725 0.5094295  3B        84  853
1805 0.2674808 0.5272989  3B        84  848
1844 0.2917435 0.5139416  OF        48  847
2013 0.3010123 0.5549457  DH         0  844
1900 0.2732327 0.5337569  OF        48  831
```

From reading the player ids, we see five similar players, in terms of career hitting statistics and position: Eddie Mathews, Mike Schmidt, Gary Sheffield, Frank Thomas, and Sammy Sosa.

8.3.4   Defining age, OBP, SLG, and OPS variables

To fit and graph hitting trajectories for a group of similar hitters, one needs to have age and OPS statistics for all seasons for each player. One complication with working with the `Batting` Lahman data base is that separate batting lines are used for batters who played with multiple teams during a season. There is a variable `stint` that gives different values (1, 2, ...) in the case of a player with multiple teams. A new function `collapse.stint` is written which collapses the counts over the `stint` variable. This function also computes the batting measures SLG, OBP, and OPS. (Recall we had earlier replaced any missing values for HBP and SF with zeros, so there will be no missing values in the calculation of the OBP and OPS variables.)

```
collapse.stint <- function(d){
  G <- sum(d$G); AB <- sum(d$AB); R <- sum(d$R)
  H <- sum(d$H); X2B <- sum(d$X2B); X3B <- sum(d$X3B)
  HR <- sum(d$HR); RBI <- sum(d$RBI); SB <- sum(d$SB)
  CS <- sum(d$CS); BB <- sum(d$BB); SH <- sum(d$SH)
  SF <- sum(d$SF); HBP <- sum(d$HBP)
  SLG <- (H - X2B - X3B - HR + 2 * X2B +
            3 * X3B + 4 * HR) / AB
  OBP <- (H + BB + HBP) / (AB + BB + HBP + SF)
  OPS <- SLG + OBP
  data.frame(G=G, AB=AB, R=R, H=H, X2B=X2B,
             X3B=X3B, HR=HR, RBI=RBI, SB=SB,
             CS=CS, BB=BB, HBP=HBP, SH=SH, SF=SF,
             SLG=SLG, OBP=OBP, OPS=OPS,
             Career.AB=d$Career.AB[1], POS=d$POS[1])
}
```

By use of the `ddply` function together with the `collapse.stint` function, we create a new version of the `Batting.2000` function where the hitting statistics for a player for a season are recorded on a single line.

```
Batting.2000 <- ddply(Batting.2000,
                   .(playerID, yearID), collapse.stint)
```

The next task is to obtain the ages for all players for all seasons. The vector `player.list` is a character vector of the player ids for all players in the `Batting.2000` data frame. Recall that a function `get.birthyear` was written in Section 8.2 to compute the MLB birth year for a particular player. By using the `sapply` function together with `get.birthyear`, we find the birth years for all players – this vector of birth years is stored in the variable `birthyears`. By use of the `merge` function, this birth year information is merged with the batting data. Now that we have birth years for all players, we can define the new variable `Age` as the difference between the season year and the birth year.

```
player.list <- as.character(unique(Batting.2000$playerID))
birthyears <- sapply(player.list, get.birthyear)
Batting.2000 <- merge(Batting.2000,
            data.frame(playerID=player.list,
                        Birthyear=birthyears))
Batting.2000$Age <- with(Batting.2000, yearID - Birthyear)
```

A small complication is that the birth year is not recorded for a few 19th century ballplayers, and so the age variable is missing for these variables. The `complete.cases` function is used to record the age records that are not missing, and the updated data frame `Batting.2000` only contains players for which the `Age` variable is available.

```
Batting.2000 <- Batting.2000[complete.cases(Batting.2000$Age), ]
```

8.3.5 Fitting and plotting trajectories

Given a group of similar players, we want to fit quadratic curves to each player and graph the trajectories in a way that facilitates comparisons. We begin by writing a function `fit.trajectory` that fits the curve for a single player with season batting statistics in data frame `d`. The output is a data frame with two variables – `Age` contains the player ages and `Fit` contains the predicted OPS using the quadratic model.

```
fit.trajectory <- function(d){
  fit <- lm(OPS ~ I(Age - 30) + I((Age - 30)^2), data = d)
  data.frame(Age = d$Age, Fit = predict(fit, Age = d$Age))
}
```

The function `plot.trajectories` graphs the trajectories for a given player together with a group of similar players. One inputs the first and last name of the player, the number of players to compare (including the one of interest), and the number of columns in the multipanel plot. This function first uses the `Master` data frame to find the player id for the player. Using the `similar` function, a vector of player ids `player.list` is found. The data frame `Batting.new` consists of the season batting statistics for only the players in the player list. The `ddply` computes the trajectories for all players (using the `fit.trajectory` function) and the ages and fitted OPS values for all players is stored in the data frame `F`. We use the `merge` function to add the players' full names to the data frame `F`. The graphing is done by use of the `ggplot2` package. The use of `geom_line` constructs trajectory curves of `Age` and `Fit` for all players. The `facet_wrap` function with the `ncol` argument places these trajectories on separate panels where the number of columns in the multipanel display is the value specified in the argument of the function.

```
plot.trajectories <- function(first, last, n.similar=5, ncol){
  require(plyr)
```

```
require(ggplot2)
get.name <- function(playerid){
  d1 <- subset(Master, playerID == playerid)
  with(d1, paste(nameFirst, nameLast))
}
player.id <- subset(Master,
              nameFirst == first & nameLast == last)$playerID
player.id <- as.character(player.id)
player.list <- as.character(similar(player.id, n.similar)$playerID)
Batting.new <- subset(Batting.2000, playerID %in% player.list)

F2 <- ddply(Batting.new, .(playerID), fit.traj)
F2 <- merge(F2,
        data.frame(playerID=player.list,
        Name=sapply(as.character(player.list), get.name)))

print(ggplot(F2, aes(Age, Fit)) + geom_line(size=1.5) +
  facet_wrap(~ Name, ncol=ncol) + theme_bw())
return(Batting.new)
}
```

Here are several examples of the use of `plot.trajectories`. Mickey Mantle's trajectory is compared with the trajectories of five similar hitters.

```
d <- plot.trajectories("Mickey", "Mantle", 6, 2)
```

We compare Derek Jeter's OPS trajectory with eight similar players.

```
d <- plot.trajectories("Derek", "Jeter", 9, 3)
```

Looking at Figure 8.3 and Figure 8.4, we see notable differences in these trajectories.

- There are players such as Eddie Mathews, Frank Thomas, Mickey Mantle, and Roberto Alomar who appeared to peak early in their careers.

- In contrast, other players such as Mike Schmidt, Craig Biggio, and Julio Franco who peaked in their 30s.

- The players also show differences in the shape of the trajectory. Johnny Damon and Julio Franco had relatively constant trajectories, and Frankie Frisch and Roberto Alomar had trajectories with high curvature.

One can summarize these trajectories by the peak age, the maximum value, and the curvature. A short function `summarize.trajectory` is written to compute these quantities for a particular fitted trajectory. The input is the data frame `d` containing the batting statistics for a player and the output is a data frame with three variables `Age.max`, `Max`, and `Curve`.

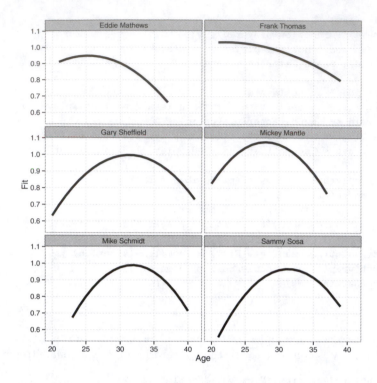

FIGURE 8.3
Fitted trajectories of OPS against age for Mickey Mantle and five similar players.

```
summarize.trajectory <- function(d){
  f <- lm(OPS ~ I(Age - 30) + I((Age - 30) ^ 2), data=d)
  b <- coef(f)
  Age.max <- round(30 - b[2] / b[3] / 2, 1)
  Max <- round(b[1] - b[2] ^ 2 / b[3] / 4, 3)
  data.frame(Age.max=Age.max, Max=Max,
             Curve=round(b[3], 5))
}
```

Recall that the output of `plot.trajectories` was a data frame containing the season batting statistics for a group of players. One can use the `ddply` function together with `summarize.trajectory` to find the summary statistics for all players. This is illustrated for Jeter and eight similar players.

```
d <- plot.trajectories("Derek", "Jeter", 9, 3)
S <- ddply(d, .(playerID), summarize.trajectory)
S
   playerID Age.max   Max    Curve
1 alomaro01    28.3 0.885 -0.00309
```

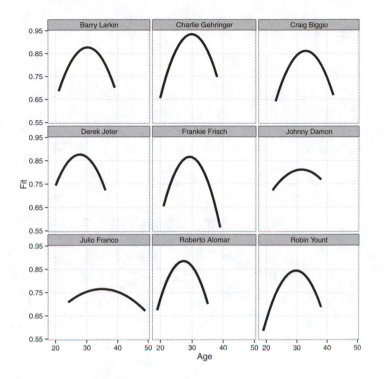

FIGURE 8.4
Fitted trajectories of OPS against age for Derek Jeter and eight similar players.

```
2 biggicr01    31.8 0.862 -0.00229
3 damonjo01    30.4 0.812 -0.00100
4 francju01    33.9 0.765 -0.00048
5 friscfr01    28.2 0.866 -0.00315
6 gehrich01    30.9 0.934 -0.00285
7 jeterde01    28.7 0.877 -0.00223
8 larkiba01    31.2 0.877 -0.00227
9 yountro01    28.7 0.844 -0.00228
```

To help understand the differences between the nine player trajectories, the `plot` function is used to construct a scatterplot of the peak ages and the curvature statistics. The `text` function is used to add player labels. (The last names of the players are found from the `Master` data frame.)

```
with(S, plot(Age.max, Curve, pch=19, cex=1.5,
          xlab="Peak Age", ylab="Curvature",
          xlim=c(27, 36), ylim=c(-0.0035, 0)))
S$lastNames <- as.character(subset(Master,
```

```
                    playerID %in% S$playerID)$nameLast)
with(S, text(Age.max, Curve, lastNames, pos=3))
```

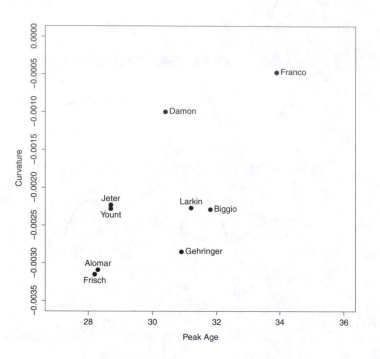

FIGURE 8.5
Scatterplot of peak age and curvature measures for Derek Jeter and eight similar players. The points are labeled by the player last names.

Looking at Figure 8.5, this clearly indicates that Alomar peaked at an early age, Franco at a late age, and Alomar and Frisch exhibited the greatest curvature, indicating they rapidly declined in performance after the peak.

8.4   General Patterns of Peak Ages

8.4.1   Computing all fitted trajectories

We have explored the hitting career trajectories of groups of similar players. How have career trajectories changed over the history of baseball? We'll focus on a player's peak age and explore how this has changed over time. Also the relationship between peak age and the number of career at-bats is explored.

For each player, the variable `playerID` contains the seasons played. We define a new variable `Midyear` defined to be the average of a player's first and last seasons. The function `ddply` is used to compute `Midyear` for all players and this new variable is added to the `Batting.2000` data frame using the `merge` function.

```
library(plyr)
midcareers <- ddply(Batting.2000, .(playerID),
                summarize,
                Midyear=(min(yearID) + max(yearID))/2)
Batting.2000 <- merge(Batting.2000, midcareers)
```

Quadratic curves to all of the career trajectories are fit by another application of the `ddply` function. A short function `coefficients.trajectory` is defined that fits the quadratic model for the season data for a particular player and returns the coefficients `A`, `B`, `C`, `Midyear`, and the player's career at-bats `Career.AB`. We apply `ddply` where `playerID` is the grouping variable and `coefficients.trajectory` is the function to be applied on each subset. The output `Beta.coef` is a data frame containing the coefficients for all players, where a row corresponds to a particular player.

```
coefficients.trajectory <- function(d){
  b <- coef(lm(OPS ~ I(Age - 30) + I((Age - 30) ^ 2), data=d))
  data.frame(A=b[1], B=b[2], C=b[3],
             Midyear=d$Midyear[1], Career.AB=d$Career.AB[1])
}
Beta.coef <- ddply(Batting.2000, .(playerID), coefficients.trajectory)
```

The estimated peak ages are computed for all players using the formula $Peak.age = 30 - B/(2C)$. The new variable `Peak.age` is added to the data frame `Beta.coef`.

```
Beta.coef$Peak.age <- with(Beta.coef, 30 - B / 2 / C)
```

8.4.2 Patterns of peak age over time

To investigate how the peak age varies over the history of baseball, we construct a scatterplot of `Peak.age` against `Midyear` by use of the `plot` function. It is difficult to see the general pattern by just looking at the scatterplot. So we use the `lowess` function to fit a smoothing curve and this curve is added to the current graph by the `lines` function. (See Figure 8.6.) (Sometimes a player's peak age estimate is infinite due to a division by zero operation, and the `is.finite` function removes those players before the application of the `lowess` function.)

```
with(Beta.coef,
    plot(Midyear, Peak.age, ylim=c(20, 40),
         xlab="Mid Career", ylab="Peak Age"))
```

```
i <- is.finite(Beta.coef$Peak.age)
with(Beta.coef,
     lines(lowess(Midyear[i], Peak.age[i]), lwd=3))
```

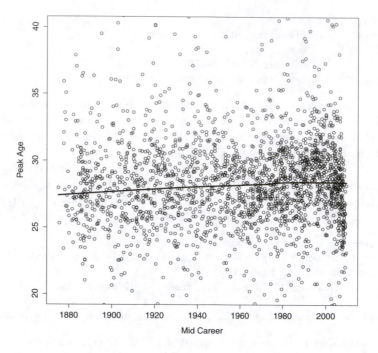

FIGURE 8.6
Scatterplot of peak age against mid career for all players with at least 2000
career at-bats.

Looking at this figure, we see a gradual increase in peak age over time. The
peak age for an average player was approximately 27 in 1880 and this average
has gradually increased to 28 from 1880 to 2000.

8.4.3   Peak age and career at-bats

Is there any relationship between a player's peak age and his career at-bats?
Using the `plot` function, we construct a graph of `Peak.age` against the logarithm (base 2) of the career at-bats variable `Career.AB`. We plot the at-bats
on a log scale, so that the points are more evenly spread out over all possible
values. Again we overlay a loess smoothing curve to see the pattern and Figure
8.7 shows the graph.

```
with(Beta.coef,
     plot(log2(Career.AB[i]), Peak.age[i], ylim=c(20, 40),
```

```
        xlab="log2 Career AB", ylab="Peak Age"))
with(Beta.coef,
    lines(lowess(log2(Career.AB[i]), Peak.age[i]), lwd=3))
```

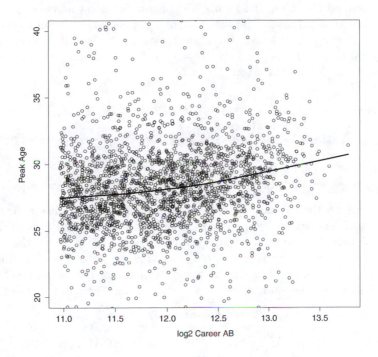

FIGURE 8.7
Scatterplot of peak age against log career AB for all players with at least 2000 career at-bats.

Here we see a clear relationship. Players with relatively short careers with 2000 career at-bats tend to peak about age 27. In contrast, players with long careers, say 9000 or more at-bats, tend to peak at ages closer to 30.

8.5 Trajectories and Fielding Position

In comparing players, we typically want to compare players at the same fielding position. The primary fielding position POS was already defined and we use this variable to compare peak ages of players categorized by position.

Suppose we consider the players whose mid-career is between 1985 and

1995. Using the `subset` function, a new data frame `Batting.2000a` is created consisting of only these players.

```
Batting.2000a <- subset(Batting.2000, Midyear >= 1985 & Midyear <= 1995)
```

A short function `trajectory.peak.pos` is written that fits a player's trajectory based on his (`Age`, `OPS`) data. The output is the estimated coefficients A, B, C, his estimated peak age `Peak.Age`, and his fielding position `Position`. By use of the `ddply` function, the trajectories are fit for all players in the `Batting2000a` data frame, and the trajectory fits for all players is stored in the data frame `Beta.estimates`.

```
trajectory.peak.pos <- function(d){
  b <- coef(lm(OPS ~ I(Age - 30) + I((Age - 30)^2), data=d))
  data.frame(A=b[1], B=b[2], C=b[3],
             Peak.Age=30 - b[2] / 2 / b[3],
             Position=d$POS[1])
}
Beta.estimates <- ddply(Batting.2000a, .(playerID), trajectory.peak.pos)
```

We focus on the primary fielding positions excluding pitcher and designated hitter. The `subset` function removes these other positions and the `Beta.estimates1$Position[ , drop=TRUE]` statement removes the levels from the factor `Position` that do not occur. The trajectory and fielding information are combined with the `Master` info by use of the `merge` function – the combined information is stored in the data frame `Beta.estimates1`.

```
Beta.estimates1 <- subset(Beta.estimates, Position %in%
                   c("1B", "2B", "3B", "SS", "C", "OF"))
Beta.estimates1$Position <- Beta.estimates1$Position[ , drop=TRUE]
Beta.estimates1 <- merge(Beta.estimates1, Master)
```

A stripchart is used to graph the peak ages of the players against the fielding position. (See Figure 8.8.) Since some of the peak age estimates are not reasonable values, the limits on the horizontal axis are set to 20 and 40.

```
stripchart(Peak.Age ~ Position, data=Beta.estimates1,
           xlim=c(20, 40), method="jitter", pch=1)
special <- with(Beta.estimates1, identify(Peak.Age, Position,
                n=5, labels=nameLast))
```

Generally, for all fielding positions, the peak ages for these 1990 players tend to fall between 27 and 32. The variability in the peak age estimates reflects the fact that hitters have different career trajectory shapes. There are three outfielders and two shortstops who seem to stand out by having a high peak age estimate. Using the `identify` function, the mouse is used to point out these unusual values and the row numbers for these players are stored in the variable `special`. The five players are Eric Davis, Gary DiSarcina, Jim Eisenreich, Alvaro Espinoza, and Tony Phillips.

A new data frame `dnew` is formed containing the hitting statistics for these

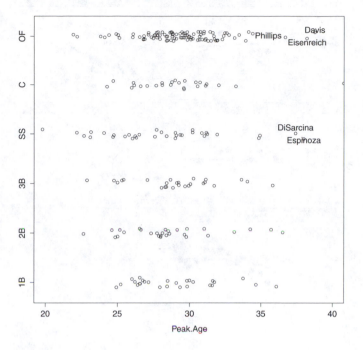

FIGURE 8.8
Peak age estimates for players with mid-career 1985–1995 graphed against fielding position.

five players with large peak age estimates, and this information is merged with the `Master` data frame. The `ggplot` function is used to graph the data together with the quadratic fits for these five players. (See Figure 8.9.) The `geom_point` function adds the points, the `stat_smooth` function adds the quadratic curves, and the `facet_wrap` function creates separate panels for the five players.

```
dnew <- subset(Batting.2000, playerID %in% Beta1$playerID[special])
dnew <- merge(dnew, Master)
ggplot(dnew, aes(Age, OPS)) + geom_point(size=4) +
  facet_wrap(~ nameLast, ncol=2) + ylim(0.4, 1.05) +
  stat_smooth(method="lm", se=FALSE, size=1.5,
              formula=y ~ poly(x, 2, raw=TRUE))  + theme_bw()
```

Looking at Figure 8.9, Davis had an unusual trajectory where he appeared to have a hitting slump in the middle of his career. DiSarcina's fitted increasing trajectory is likely caused by the one large OPS value towards the end of his career. The general impression is that Eisenreich, Espinoza, and Phillips had pretty consistent level OPS values through their careers.

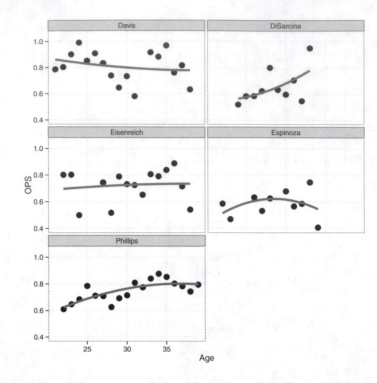

FIGURE 8.9
OPS trajectory graphs for five players with unusually large peak age estimates.

8.6   Further Reading

James (1982) wrote an essay on "Looking For the Prime"; based on a statistical study, he came to the conclusion that batters tend to peak at age 27. Berry et al. (1999) give a general discussion of career trajectories of athletes from hockey, baseball, and golf. Chapter 11 of Albert and Bennett (2003) considers the career trajectories of the home run rates of nine great historical sluggers. Albert (2002, 2009) discuss general patterns of trajectories of hitters and pitchers in baseball history, and Fair (2008) performs an extensive analysis of baseball career trajectories based on quadratic models. Albert and Rizzo (2012), Chapter 7, give illustrations of regression modeling using R.

8.7 Exercises

1. (**Career Trajectory of Willie Mays**)

   (a) Use the `gets.stats` function to extract the hitting data for Willie Mays for all of his seasons in his career.

   (b) Construct a scatterplot of Mays' OPS season values against his age.

   (c) Fit a quadratic function to Mays' career trajectory. Based on this model, estimate Mays' peak age and his estimated largest OPS value based on the fit.

2. (**Comparing Trajectories**)

   (a) Using James' similarity score measure (function `similar`), find the five hitters with hitting statistics most similar to Willie Mays.

   (b) Fit quadratic functions to the (Age, OPS) data for Mays and the five similar hitters. Display the six fitted trajectories on a single panel.

   (c) Based on your graph, describe the differences between the six player trajectories. Which player had the smallest peak age?

3. (**Comparing Trajectories of the Career Hits Leaders**)

   (a) Find the batters who have had at least 3200 career hits.

   (b) Fit the quadratic functions to the (Age, AVG) data for this group of hitters, where AVG is the batting average. Display the fitted trajectories on a single panel.

   (c) On the basis of your work, which player was the most consistent hitter on average? Explain how you measured consistency on the basis of the fitted trajectory.

4. (**Comparing Trajectories of Home Run Hitters**)

   (a) Find the ten players in baseball history who have had the most career home runs.

   (b) Fit the quadratic functions to the home run rates of the ten players, where $HR.RATE = HR/AB$. Display the fitted trajectories on a single panel.

   (c) On the basis of your work, which player had the highest estimated home run rate at his peak? Which player among the ten had the smallest peak home run rate?

   (d) Do any of the players have unusual career trajectory shapes? Is there any possible explanation for these unusual shapes?

5. (**Peak Ages in the History of Baseball**)

(a) Find all the players who entered baseball between 1940 and 1945 with at least 2000 career at-bats.

(b) Find all the players who entered baseball between 1970 and 1975 with at least 2000 career at-bats.

(c) By fitting quadratic functions to the (Age, OPS) data, estimate the peak ages for all players in parts (a) and (b).

(d) By comparing the peak ages of the 1940s players with the peak ages of the 1970s players, can you make any conclusions about how the peak ages have changed in this 30-year period?

9

Simulation

CONTENTS

9.1	Introduction		211
9.2	Simulating a Half Inning		212
	9.2.1	Markov chains	212
	9.2.2	Review of work in runs expectancy	213
	9.2.3	Computing the transition probabilities	215
	9.2.4	Simulating the Markov chain	216
	9.2.5	Beyond runs expectancy	219
	9.2.6	Transition probabilities for individual teams	220
9.3	Simulating a Baseball Season		223
	9.3.1	The Bradley-Terry model	223
	9.3.2	Making up a schedule	224
	9.3.3	Simulating talents and computing win probabilities	225
	9.3.4	Simulating the regular season	225
	9.3.5	Simulating the post-season	226
	9.3.6	Function to simulate one season	227
	9.3.7	Simulating many seasons	228
9.4	Further Reading		231
9.5	Exercises		232

9.1 Introduction

A baseball season consists of a collection of games between teams, where each game consists of nine innings, and a half-inning consists of a sequence of plate appearances. Because of this clean structure, the sport can be represented by relatively simple probability models. Simulations from these models are helpful in understanding different characteristics of the game.

One attractive aspect of the R system is its ability to simulate from a wide variety of probability distributions. In this chapter, we illustrate the use of R functions to simulate a game consisting of a large number of plate appearances. Also, R is used to simulate the game-to-game team competition of teams during an entire season.

Section 9.2 focuses on simulating the events in a baseball half-inning using a special probability model called a Markov chain. The runners on base and the number of outs define a state and this probability model describes movements between states until one reaches three outs. The movement or transition probabilities are found using actual data from the 2011 season. By simulating many half-innings using this model, one gets a basic understanding of the pattern of run scoring.

Section 9.3 describes a simulation of an entire baseball season using the Bradley-Terry probability model. Teams are assigned talents from a bell-shaped (normal) distribution and a season of baseball games is played using win probabilities based on the talents. By simulating many seasons, one learns about the relationship between a team's talent and its performance in a 162-game season. We describe simulating the post-season series and assess the probability that the "best" team, that is, the team with the best ability actually wins the World Series.

9.2   Simulating a Half Inning

9.2.1   Markov chains

A Markov chain is a special type of probability model useful for describing movement between locations, called *states*. In the baseball context, a state is viewed as a description of the runners on base and the number of outs in an inning. Each of the three bases can be occupied by a runner or not, and so there are $2 \times 2 \times 2 = 8$ possible runner situations. Since there are three possible numbers of outs (0, 1, or 2), there are $8 \times 3 = 24$ possible runner and outs states. If we include the 3 outs state, there are a total of 25 possible states during a half-inning of baseball.

In a Markov chain, a matrix of *transition probabilities* is used to describe how one moves between the different states. For example, suppose that there are currently runners on first and second with one out. Based on the outcome of the plate appearance, the state can change. For example, the batter may hit a single; the runner on second scores and the runner on first moves to third. In this case, the new state is runners on first and third with one out. Or maybe the batter will strike out, and the new state is runners on first and second with two outs. By looking at a specific row in the transition probability matrix, one learns about the probability of moving to first and third with one out, or moving to first and second with two outs, or any other possible state.

In a Markov chain, there are two types of states – transition states and absorbing states. Once one moves into an absorbing state, one remains there and can't return to other transition states. In a half-inning of baseball, since the inning is over when there are 3 outs, this 3-outs state acts as an absorbing state.

There are some special assumptions in a Markov chain model. We assume that the probability of moving to a new state only depends on the current state. So any baseball events that happened before the current runners and outs situation are not relevant in finding the probabilities. In other words, this model assumes there is not a momentum effect in batting through an inning. Also we are assuming that the probabilities of these movements are the same for all teams, against all pitchers, and for all innings during a game. Clearly, this assumption that all teams are average is not realistic, but we will address this issue in one of the sections of this chapter.

There are several attractive aspects of using a Markov chain to model a half-inning of baseball. First, the construction of the transition probability matrix is easily done with 2011 season data using computations from the runs expectancy chapter. One can use the model to play many half-innings of baseball and the run scoring patterns that are found resemble the actual run scoring of actual MLB baseball. Last, there are special properties of Markov chains that simplify some interesting calculations, such as the number of players who come to bat during an inning.

9.2.2 Review of work in runs expectancy

To construct the transition matrix for the Markov chain, one needs to know the frequencies of transitions from the different runners/outs states to other possible runners/outs states. One obtains these frequencies using the Retrosheet play-by-play data from a particular season, and we review the work from Chapter 5.

We begin by reading in play-by-play data for the 2011 season, creating the data frame `data2011`.

```
data2011 <- read.csv("all2011.csv", header=FALSE)
fields <- read.csv("fields.csv")
names(data2011) <- fields[, "Header"]
```

The variable `HALF.INNING` is created which creates a unique identification for each half-inning in each baseball game. The new variable `RUNS.SCORED` gives the number of runs scored in each play.

```
data2011$HALF.INNING <- with(data2011,
        paste(GAME_ID, INN_CT, BAT_HOME_ID))
data2011$RUNS.SCORED <- with(data2011, (BAT_DEST_ID > 3) +
  (RUN1_DEST_ID > 3) + (RUN2_DEST_ID > 3) + (RUN3_DEST_ID > 3))
```

The function `get.state` defines a state variable based on the runners on the three bases (`runner1`, `runner2`, `runner3`) and the number of outs (`outs`). Using this function, the new variable `STATE` is computed which gives the runner locations and the number of outs at the beginning of each play. The variable `NEW.STATE` contains the runners and outs information at the conclusion of the play.

```
get.state <- function(runner1, runner2, runner3, outs){
  runners <- paste(runner1, runner2, runner3, sep="")
  paste(runners, outs)
}


RUNNER1 <- ifelse(as.character(data2011[,"BASE1_RUN_ID"])=="", 0, 1)
RUNNER2 <- ifelse(as.character(data2011[,"BASE2_RUN_ID"])=="", 0, 1)
RUNNER3 <- ifelse(as.character(data2011[,"BASE3_RUN_ID"])=="", 0, 1)
data2011$STATE <- get.state(RUNNER1, RUNNER2, RUNNER3,
                            data2011$OUTS_CT)

NRUNNER1 <- with(data2011, as.numeric(RUN1_DEST_ID==1 |
                          BAT_DEST_ID==1))
NRUNNER2 <- with(data2011, as.numeric(RUN1_DEST_ID==2 |
                  RUN2_DEST_ID==2 | BAT_DEST_ID==2))
NRUNNER3 <- with(data2011, as.numeric(RUN1_DEST_ID==3 |
                  RUN2_DEST_ID==3 | RUN3_DEST_ID==3 | BAT_DEST_ID==3))
NOUTS <- with(data2011, OUTS_CT + EVENT_OUTS_CT)
data2011$NEW.STATE <- get.state(NRUNNER1, NRUNNER2, NRUNNER3, NOUTS)
```

By using the `subset` function, we focus on plays where there is a change in the state or in the number of runs scored. By another application of `subset`, we restrict attention to complete innings where there are three outs – the new data frame is named `data2011C`. Last, by use of the `BAT_EVENT_FL` variable, we only consider plays where there is a batting event. So non-batting plays such as steals, caught stealing, wild pitches, and passed balls are ignored. There is obviously some consequence of removing these non-batting plays from the viewpoint of run production, and this issue is discussed later in this chapter.

```
data2011 <- subset(data2011, (STATE != NEW.STATE) | (RUNS.SCORED > 0))


library(plyr)
data.outs <- ddply(data2011, .(HALF.INNING), summarize,
                   Outs.Inning=sum(EVENT_OUTS_CT))
data2011 <- merge(data2011, data.outs)
data2011C <- subset(data2011, Outs.Inning == 3)

data2011C <- subset(data2011, BAT_EVENT_FL == TRUE)
```

In our definition of the `NEW.STATE` variable, we recorded the runner locations when there were three outs. The runner locations don't matter, so we recode `NEW.STATE` to the value "3" when the number of outs is equal to 3.

```
library(car)
data2011C$NEW.STATE <- recode(data2011C$NEW.STATE,
            "c('000 3', '100 3', '010 3', '001 3',
               '110 3', '101 3', '011 3', '111 3')='3'")
```

9.2.3 Computing the transition probabilities

Now that the `STATE` and `NEW.STATE` variables are defined, one can compute the frequencies of all possible transitions between states by use of the `table` function – the matrix of counts is `T.matrix`. There are 24 possible values of the beginning state `STATE`, and 25 values of the final state `NEW.STATE` including the 3-outs state.

```
T.matrix <- with(data2011C, table(STATE, NEW.STATE))
```

This matrix can be converted to a probability matrix by use of the `prop.table` function; the resulting matrix is denoted by `P.matrix`.

```
P.matrix <- prop.table(T.matrix, 1)
```

We add a row to this transition probability matrix corresponding to transitions from the 3-out state. When the inning reaches 3-outs, then it stays at 3-outs, so the probability of staying in this state is 1.

```
P.matrix <- rbind(P.matrix, c(rep(0, 24), 1))
```

To better understand this transition matrix, the transition probabilities starting at the "000 0" state, no runners and no outs are displayed. (Only the positive probabilities are shown and the `data.frame` function is used to display the probabilities vertically.) The most likely transitions are to the "no runners, one out" state with probability 0.677 and to the "runner on first, no outs" state with probability 0.240. The probability of moving from the "000 0" state to the "000 0" state is 0.027; in other words, the chance of a home run with no runners on with no outs is 0.027.

```
P1 <- round(P.matrix["000 0", ], 3)
data.frame(Prob=P1[P1 > 0])
        Prob
000 0 0.027
000 1 0.677
001 0 0.006
010 0 0.050
100 0 0.240
```

Let's contrast this with the possible transitions starting from the "010 2" state, runner on second with two outs. The most likely transitions are "3 outs" (probability 0.640), "runners on first and second with two outs" (probability 0.163), and "runner on first with 2 outs" (probability 0.083).

```
P2 <- round(P.matrix["010 2", ], 3)
data.frame(Prob=P2[P2 > 0])
        Prob
000 2 0.020
001 2 0.006
010 2 0.055
```

```
100 2 0.083
101 2 0.034
110 2 0.163
3     0.640
```

9.2.4   Simulating the Markov chain

One can simulate this Markov chain model a large number of times to obtain
the distribution of runs scored in a half-inning of 2011 baseball. The first
step is to construct a matrix giving the runs scored in all possible transitions
between states. Let $N_{runners}$ denote the number of runners in a state and O
denote the number of outs. For a batting play, the number of runs scored is
equal to

$$RUNS = (N_{runners}^{(b)} + O^{(b)} + 1) - (N_{runners}^{(a)} + O^{(a)}).$$

In other words, the runs scored is the sum of runners and outs before (b)
the play minus the sum of runners and outs after (a) the play plus one. For
example, suppose there are runners on first and second with one outs, and
after the play, there is a runner on second with two outs. The number of runs
scored is equal to

$$RUNS = (2 + 1 + 1) - (1 + 2) = 1.$$

A new function `count.runners.outs` is defined which takes a state as input and returns the sum of the number of runners and outs. This function is
applied across all the possible states (using the `sapply` function) and the corresponding sums are stored in the vector `runners.outs`. The `outer` function
with the "-" operation performs the RUNS calculation for all possible pairs
of states and the resulting matrix is stored in the variable `R`. If one inspects
the matrix `R`, one will notice some negative values and some strange large
positive values. But this is not a concern since the corresponding transitions,
for example a movement between a "000 0" state and a "000 2" state in one
batting play, are not possible. An additional column of zeros is added to this
run matrix by use of the `cbind` function.

```
count.runners.outs <- function(s)
  sum(as.numeric(strsplit(s,"")[[1]]), na.rm=TRUE)
runners.outs <- sapply(dimnames(T.matrix)[[1]], count.runners.outs)[-25]
R <- outer(runners.outs + 1, runners.outs, FUN="-")
dimnames(R)[[1]] <- dimnames(T.matrix)[[1]][-25]
dimnames(R)[[2]] <- dimnames(T.matrix)[[1]][-25]
R <- cbind(R, rep(0, 24))
```

We are now ready to simulate a half-inning of baseball using a new function
`simulate.half.inning`. The inputs are the probability transition matrix P,
the run matrix R, and the starting state s (an integer between 1 and 24). The
output is the number of runs scored in the half-inning.

```
simulate.half.inning <- function(P, R, start=1){
  s <- start; path <- NULL; runs <- 0
  while(s < 25){
    s.new <- sample(1:25, 1, prob=P[s, ])
    path <- c(path, s.new)
    runs <- runs + R[s, s.new]
    s <- s.new
  }
  runs
}
```

There are two key statements in this simulation. If the current state is s, the function `sample` will simulate a new state using the s row in the transition matrix P – the new state is denoted s.new. The total number of runs scored in the inning is updated using the value in the s row and the s.new column of the runs matrix R.

Using the `replicate` function, one can simulate a large number of half-innings of baseball. In the below code, we simulate 10,000 half-innings starting with no runners and no outs (state 1), collecting the runs scored in the vector RUNS.

```
RUNS <- replicate(10000, simulate.half.inning(T.matrix, R))
```

To find the possible runs scored in a half-inning, the `table` function is used to tabulate the values in RUNS.

```
table(RUNS)
RUNS
   0    1    2    3    4    5    6    7    8    9   10
7483 1334  659  312  133   44   22    8    1    3    1
```

In our 10,000 simulations, five or more runs scored in $44 + 22 + 8 + 1 + 3 + 1 = 79$ half-innings, so the chance of scoring five or more runs would be 79 / 10,000 = 0.0079. This calculation can be checked using the `sum` function.

```
sum(RUNS[RUNS >= 5]) / 10000
[1] 0.0079
```

The mean number of runs scored is computed by applying the `mean` function on RUNS.

```
mean(RUNS)
[1] 0.4584
```

Over the 10,000 half-innings, an average of 0.4584 runs were scored.

To understand the runs potential of different runners and outs situations, one can repeat this simulation procedure for other starting states. A function RUNS.j is written to compute the mean number of runs scored starting with state j. By use of the `sapply` function, we apply the function RUNS.j over all of the possible starting states 1 through 24. The output is a vector of mean runs scored stored in the variable Runs.Expectancy.

```
RUNS.j <- function(j){
  mean(replicate(10000, simulate.half.inning(T.matrix, R, j)))
}
Runs.Expectancy <- sapply(1:24, RUNS.j)
Runs.Expectancy <- t(round(matrix(Runs.Expectancy, 3, 8), 2))
dimnames(Runs.Expectancy)[[2]] <- c("0 outs", "1 out", "2 outs")
dimnames(Runs.Expectancy)[[1]] <- c("000", "001", "010", "011", "100",
                       "101",  "110", "111")
Runs.Expectancy
```

	0 outs	1 out	2 outs
000	0.45	0.26	0.09
001	1.36	0.91	0.32
010	1.08	0.63	0.30
011	1.90	1.33	0.50
100	0.82	0.49	0.21
101	1.73	1.13	0.47
110	1.43	0.86	0.40
111	2.23	1.52	0.69

Recall that our simulation model is based only on batting plays. To understand the effect of non-batting plays (stealing, caught stealing, wild pitches, etc.) on run scoring, we compare this runs expectancy matrix with the one found in Chapter 5 using all batting and non-batting plays. A new matrix Runs is created with the earlier matrix and we compute the difference Runs – Runs.Expectancy – this is the contribution of non-batting plays to the average number of runs scored.

```
Runs <- matrix(
  c(0.47, 0.25, 0.10, 1.45, 0.94, 0.32,
    1.06, 0.65, 0.31, 1.93, 1.34, 0.54,
    0.84, 0.50, 0.22, 1.75, 1.15, 0.49,
    1.41, 0.87, 0.42, 2.17, 1.47, 0.76),
  8, 3, byrow=TRUE)

Runs - Runs.Expectancy
```

	0 outs	1 out	2 outs
000	0.02	-0.01	0.01
001	0.09	0.03	0.00
010	-0.02	0.02	0.01
011	0.03	0.01	0.04
100	0.02	0.01	0.01
101	0.02	0.02	0.02
110	-0.02	0.01	0.02
111	-0.06	-0.05	0.07

Note that most of the values of the difference are positive, indicating that these non-batting plays generally do create runs. The three largest values are 0.09, 0.07, and 0.04 corresponding to the "001, 0 outs", "111, 2 outs", and "011, 2 outs" situations. These positive values make sense since these are all situations with a runner on third who can score with a wild pitch or passed ball.

9.2.5 Beyond runs expectancy

By using properties of Markov chains, it is straightforward to use the transition probability matrix to learn more about the movement through the runners/outs states.

By multiplying the probability matrix `P.matrix` by itself three times, we can learn about the state of the inning after three plate appearances. In R, matrix multiplication is indicated by the `%*%` symbol. The result is stored in the matrix `P.matrix.3`.

```
P.matrix.3 <- P.matrix %*% P.matrix %*% P.matrix
```

The first row of `P.matrix.3` gives the probabilities of being in each of the 25 states after three hitters starting at the "000 0" state. We round these values to three decimal places, sort from largest to smallest, and display the largest values.

```
sorted.P <- sort(round(P.matrix.3["000 0", ], 3), decreasing=TRUE)
head(data.frame(Prob=sorted.P))
        Prob
3       0.369
100 2 0.241
110 1 0.086
010 2 0.083
000 2 0.045
001 2 0.030
```

After three PAs, the most likely outcomes are three outs (probability 0.369), runner on first with 2 outs (probability 0.241), and runners on first and second with one out (probability 0.086).

It is also easy to learn about the number of visits to all runner-outs states. Define the matrix `Q` to be the 24-by-24 submatrix found from the transition matrix by removing the last row and column (the three outs state). By subtracting the matrix `Q` from the identity matrix and taking the inverse of the result, we obtain the fundamental matrix `N` of an absorbing Markov chain. (The `diag` function is used to construct the identity matrix and the function `solve` takes the matrix inverse.)

```
Q <- P.matrix[-25, -25]
N <- solve(diag(rep(1, 24)) - Q)
```

To understand the fundamental matrix, the beginning entries of the first row of the matrix are displayed.

```
N.0000 <- round(N["000 0", ], 2)
head(data.frame(N=N.0000))
```

```
        N
000 0 1.04
000 1 0.74
000 2 0.58
001 0 0.01
001 1 0.03
001 2 0.05
```

Starting at the beginning of the inning (the "000 0" state), the average number of times the inning will be in the "000 0" state is 1.04, the average number of times in the "000 1" state is 0.74, the average number of times in the "000 2" state is 0.58, and so on. By using the sum function, we find the average number of states that are visited.

```
sum(N.0000)
[1] 4.28
```

In other words, the average number of plate appearances in a half-inning (before three outs) is 4.28.

We can compute the average number of batting plays until three outs for all starting states by multiplying the fundamental matrix N by a column vector of ones. The vector of average number of plays is stored in the variable Length and eight values of this vector are displayed.

```
Length <- round(t(N %*% rep(1, 24)), 2)
data.frame(L=Length[1, 1:8])
        L
000 0 4.28
000 1 2.88
000 2 1.47
001 0 4.37
001 1 2.97
001 2 1.51
010 0 4.33
010 1 2.95
```

This tells us the length of the remainder of the inning, on average, starting with each possible state. For example, starting at the bases empty, one out state, we expect on average to have 2.88 more batters. In contrast, with a runner on third with two outs, we expect to have 1.51 more batters.

9.2.6   Transition probabilities for individual teams

The transition probability matrix describes movements between states for an average team. Certainly, these probabilities will vary for teams of different batting abilities, and the probabilities will also vary against teams of different pitching abilities. We focus on different batting teams and discuss how to obtain good estimates of the transition probabilities for all teams.

To get the relevant data, a new variable BATTING.TEAM needs to be defined that gives the batting team in each half-inning. By use of the `substr` function, we define the home team variable `HOME_TEAM_ID`, and an `ifelse` function is used to define the batting team.

```
data2011C$HOME_TEAM_ID <- with(data2011C, substr(GAME_ID, 1, 3))
data2011C$BATTING.TEAM <- with(data2011C,
        ifelse(BAT_HOME_ID == 0,
               as.character(AWAY_TEAM_ID),
               as.character(HOME_TEAM_ID)))
```

By use of the `table` function, a three-way table `Team.T` is constructed giving the counts of each team in the transitions from the current to new states.

```
Team.T <- with(data2011C, table(BATTING.TEAM, STATE, NEW.STATE))
```

For example, the matrix `Team.T['ANA', , ]` gives the transition counts for Anaheim in the 2011 season.

If one is interested in comparing run productions for different batting teams, it is necessary to make some adjustments to the team transition probability matrices to get realistic predictions of performance. To illustrate the problem, we focus on transitions from the "100 2" state. The transition counts are stored in the variable `Team.T.S` and a few rows of this table are displayed. for six of the teams.

```
d.state <- subset(data2011C, STATE == '100 2')
Team.T.S <- with(d.state, table(BATTING.TEAM, NEW.STATE))
Team.T.S
```

	NEW.STATE							
BATTING.TEAM	000 2	001 2	010 2	011 2	100 2	101 2	110 2	3
ANA	11	3	7	8	0	16	56	253
ARI	11	4	13	2	0	15	73	240
ATL	7	2	4	7	0	23	68	273
...								
TEX	12	5	16	6	1	20	67	268
TOR	7	3	9	10	1	18	51	269
WAS	9	1	5	10	0	25	61	243

For some of the less common transitions, there is much variability in the counts across teams and this causes the corresponding team transition probabilities to be unreliable. If p^{TEAM} represents the team's transition probabilities for a particular team, and p^{ALL} are the average transition probabilities, then a better estimate at the team's probabilities has the form

$$p^{EST} = \frac{n}{n+K}p^{TEAM} + \frac{K}{n+K}p^{ALL},$$

where n is the number of transitions for the team and K is a smoothing count. The description of the methodology is beyond the level of this book, but in this case a smoothing count of $K = 1274$ leads to a good estimate at the team's transition probabilities. (The choice of K depends on the starting state.)

This method is illustrated for Washington's transition counts starting from the "100 2" state. The transition counts are stored in WAS.Trans, the total number of transitions are in WAS.n, and the team transition proportions in P.WAS. The transition counts for all teams are stored in ALL.Trans, and the overall transition proportions are in P.ALL. The smoothing count of 1274 is stored in K.

```
WAS.Trans <- Team.T.S["WAS", ]
WAS.n <- sum(WAS.Trans)
P.WAS <- WAS.Trans / WAS.n

ALL.Trans <- with(subset(data2011C, STATE == '100 2'),
              table(NEW.STATE))
P.ALL <- ALL.Trans / sum(ALL.Trans)
K <- 1274
```

The improved estimate at Washington's transition proportions is computed using the formula and stored in P.EST. The three sets of proportions (Washington, overall, and improved) are displayed in a data frame.

```
P.EST <- WAS.n / (K + WAS.n) * P.WAS + K / (K + WAS.n) * P.ALL
data.frame(WAS=round(P.WAS, 4),
          ALL=round(c(P.ALL), 4),
          EST=round(c(P.EST), 4))
        WAS     ALL     EST
000 2 0.0254 0.0254 0.0254
001 2 0.0028 0.0073 0.0063
010 2 0.0141 0.0242 0.0220
011 2 0.0282 0.0209 0.0225
100 2 0.0000 0.0005 0.0004
101 2 0.0706 0.0504 0.0548
110 2 0.1723 0.1885 0.1850
3     0.6864 0.6827 0.6835
```

Note that the improved transition proportions are a compromise between the team's proportions and the overall values. For example, for a transition from the state "100 2" to "010 2", the Washington value is 0.0141, the overall value is 0.0242, and the improved value 0.0220 falls between the Washington and overall values. This method is especially helpful for transitions such as "100 2" to "100 2" which did not occur for Washington in this season but we know there is a positive chance of these transitions happening in the future.

This smoothing method can be applied for all teams and all rows of the transition matrix to obtain improved estimates at teams' probability transition matrices. With the team transition matrices computed in this way, one can explore the run-scoring behavior of individual batting teams.

9.3   Simulating a Baseball Season

9.3.1   The Bradley-Terry model

An attractive method of modeling paired comparison data such as baseball games is the Bradley-Terry model. This modeling and simulation is illustrated for the 1968 baseball season where the regular season and playoff system had a relatively simple structure. It is straightforward to adapt these methods to the present baseball season with a more complicated schedule and playoff system.

In 1968, there were 20 teams, 10 in the National League and 10 in the American League. Suppose each team has a talent or ability to win a game. The talents for the 20 teams are represented by the values $T_1, ..., T_{20}$. We assume that the talents are distributed from a normal curve model with mean 0 and standard deviation s_T. A team of average ability would have a talent value close to zero, "good" teams would have positive talents, and bad teams would have negative talents. Suppose team A plays team B in a single game. By the Bradley-Terry model, the probability team A wins the game is given by the logistic function

$$P(A\,wins) = \frac{\exp(T_A)}{\exp(T_A) + \exp(T_B)}.$$

This model is closely related to the log5 method developed by Bill James in his *Baseball Abstract* books in the 1980s. If P_A and P_B are the winning percentages of teams A and B, then James' formula is given by

$$P(A\,wins) = \frac{P_A/(1 - P_A)}{P_A/(1 - P_A) + P_B/(1 - P_B)}.$$

Comparing the two formulas, one sees that the log5 method is a special case of the Bradley-Terry model where a team's talent T is set equal to the log odds of winning $\log(P/(1 - P))$. A team with a talent $T = 0$ will win (in the long run) half of its games ($P = 0.5$). In contrast, a team with talent $T = 0.2$ will win (using the log 5 values) approximately 55% of its games and a team with talent $T = -0.2$ will win 45% of its games.

Using this model, one can simulate a baseball season as follows.

1. Construct the 1968 baseball schedule. In this season, each of the 10 teams in each league play each other team in the same league 18 games, where 9 games are played in each team's ballpark. (There was no interleague play in 1968.)

2. Simulate 20 talents from a normal distribution with mean 0 and standard deviation s_T. The value of s_T is chosen so that the simulated season winning percentages from this model resemble the actual winning percentages during this season.

3. Using the probability formula and the talent values, one computes the probabilities that the home team wins all games. By a series of coin flips with these probabilities, one determines the winners of all games.

4. Determine the winner of each league (ties need to be broken by some random mechanism) and play a best-of-seven World Series using winning probabilities computed using the Bradley-Terry formula and the two talent numbers.

9.3.2   Making up a schedule

The first step in the simulation is to construct the schedule of games. A short function `make.schedule` is written to help with this task. The inputs are the vector of team names `teams` and the number of games k that will be played between two teams in the first team's home park. The output is a data frame where each row corresponds to a game and `Home` and `Visitor` give the names of the home and visiting teams. The `gl` function is helpful for generating a factor which repeats each team label a particular number of replications, and the `rep` function generates repeated copies of a vector.

```
make.schedule <- function(teams, k){
    n.teams <- length(teams)
    Home <- rep(gl(n.teams, n.teams, length=n.teams ^ 2,
                   labels=teams), k)
    Visitor <- rep(gl(n.teams, 1, length=n.teams ^ 2, labels=teams), k)
    schedule <- data.frame(Home=Home, Visitor=Visitor)
    subset(schedule, Home != Visitor)
}
```

This function is used to construct the schedule for the 1968 season. Two vectors `NL` and `AL` are constructed containing abbreviations for the National League and American League teams. We apply the function `make.schedule` twice, once for each league, using k = 9 since one team hosts another team nine games. The `rbind` is used to paste together the NL and AL schedules, creating the data frame `schedule`.

```
NL <- c("ATL", "CHN", "CIN", "HOU", "LAN", "NYN", "PHI",
          "PIT", "SFN", "SLN")
AL <- c("BAL", "BOS", "CAL", "CHA", "CLE", "DET", "MIN",
          "NYA", "OAK", "WS2")
teams <- c(NL, AL)
league <- c(rep(1, 10), rep(2, 10))
schedule <- rbind(make.schedule(NL, 9),
                   make.schedule(AL, 9))
```

9.3.3 Simulating talents and computing win probabilities

The next step to compute the win probabilities for all of the games in the season schedule. The team talents are assumed to come from a normal distribution with mean 0 and standard deviation `s.talent`, which we assign `s.talent = 0.20`. (Recall that this value of the standard deviation is chosen so that the season team win percentages generated from the model resemble the actual team win percentages.) The talents are simulated using the function `rnorm` and a data frame `TAL` is created that assigns the talents to the 20 teams. By use of several applications of the `merge` function, we add the team talents to the `schedule` data frame – the new data frame is called `SCH`. Last, since we have the talents for the home and visiting teams for all games, the Bradley-Terry model is applied to compute home team winning probabilities for all games – these probabilities are stored in the variable `prob.Home`.

```
s.talent <- 0.20
talents <- rnorm(20, 0, s.talent)
TAL <- data.frame(Team=teams, League=league, Talent=talents)
SCH <- merge(schedule, TAL, by.x="Home", by.y="Team")
names(SCH)[4] <- "Talent.Home"
SCH <- merge(SCH, TAL, by.x="Visitor", by.y="Team")
names(SCH)[6] <- "Talent.Visitor"
SCH$prob.Home <- with(SCH,
        exp(Talent.Home) / (exp(Talent.Home) + exp(Talent.Visitor)))
```

The first six rows of the data frame `SCH` are displayed, where one sees the games scheduled, the talents of the home and away teams, and the probability that the home team wins the matchup.

```
head(SCH)
  Visitor Home League.x Talent.Home League.y Talent.Visitor prob.Home
1     ATL  PHI        1 -0.02757542        1    -0.06858703 0.5102515
2     ATL  PIT        1  0.14765145        1    -0.06858703 0.5538500
3     ATL  SFN        1 -0.07264467        1    -0.06858703 0.4989856
4     ATL  NYN        1 -0.20925903        1    -0.06858703 0.4648899
5     ATL  HOU        1 -0.19522186        1    -0.06858703 0.4683835
6     ATL  SLN        1 -0.09780258        1    -0.06858703 0.4926966
```

9.3.4 Simulating the regular season

To simulate an entire season of games, we are doing a series of coin flips, where the probability the home team wins depends on the winning probability. The function `rbinom` performs the coin flips for the 1620 scheduled games – the outcomes are a sequence of 0s and 1s. By use of the `ifelse` function, we define the `winner` variable to be the `Home` team if the outcome is 1, and the `Visitor` otherwise.

```
SCH$outcome <- with(SCH, rbinom(nrow(SCH), 1, prob.Home))
SCH$winner <- with(SCH, ifelse(outcome, as.character(Home),
                    as.character(Visitor)))
```

The teams, home win probabilities, and outcomes of the first six games are displayed.

```
head(SCH[, c("Visitor", "Home", "prob.Home", "outcome","winner")])
  Visitor Home prob.Home outcome winner
1     ATL  PHI 0.5102515       0    ATL
2     ATL  PIT 0.5538500       1    PIT
3     ATL  SFN 0.4989856       1    SFN
4     ATL  NYN 0.4648899       1    NYN
5     ATL  HOU 0.4683835       1    HOU
6     ATL  SLN 0.4926966       1    SLN
```

How did the teams perform during this particular simulated season? Using the `table` function, we find the number of wins for all teams. This information is collected together with the team names in the data frame `WIN`, and using the `merge` function the season results are combined with the team talents to create the data frame `RESULTS`.

```
wins <- table(SCH$winner)
WIN <- data.frame(Team=names(wins), Wins=as.numeric(wins))
RESULTS <- merge(TAL, WIN)
```

9.3.5   Simulating the post-season

After the regular season, one simulates the post-season series. A function `win.league` is written that simulates a league championship – the inputs are the data frame `RR` of teams and win totals and the league number (1 for National and 2 for American). The function first identifies the teams that have the largest number of wins. If one team has the maximum number, then an indicator variable `Winner.Lg` is created which is 1 for that particular team. If there is a tie in win totals for two or more teams, then we randomly choose one of the teams to be the pennant winner (using the random multinomial function `rmultinom`) where the winning probabilities are proportional to the exponentials of the talents.

```
win.league <- function(RR, league){
    wins <- RR$Wins * (RR$League == league)
    MAX <- max(wins)
    if(sum(wins == MAX) > 1){
      prob <- exp(RR$Talent) * (wins == MAX)
      outcome <- c(rmultinom(1, 1, prob))
      RR$Winner.Lg <- RR$Winner.Lg + outcome
    }
    if(sum(wins == MAX) == 1){
```

```
        RR$Winner.Lg <- RR$Winner.Lg + as.numeric(wins == MAX)}
     RR
}
```

To simulate the post-season, we initialize two new variables `Winner.Lg` and `Winner.WS` – these are indicators for the league champions and the World Series winners. By two applications of `win.league`, we find the winners of each league. The World Series is simulated by flipping a coin seven times, where the win probabilities are proportional to exp(Talent). The vector `winner` indicates the team winning a majority of the games.

```
RESULTS$Winner.Lg <- 0
RESULTS$Winner.WS <- 0;
for(j in 1:2)
    RESULTS <- win.league(RESULTS, j)
teams <- (1:20)[RESULTS$Winner.Lg == 1]
outcome <- c(rmultinom(1, 7, exp(RESULTS$Talent)[teams]))
winner <- teams[1] * (diff(outcome) < 0) + teams[2] * (diff(outcome) > 0)
RESULTS$Winner.WS[winner] <- 1
```

9.3.6 Function to simulate one season

It is convenient to place all of these commands including the functions `make.schedule` and `win.league` in a single function `one.simulation.68`. The only input is the standard deviation `s.talent` that describes the spread of the normal talent distribution. The output is a data frame containing the teams, talents, number of season wins, and success in the post-season. We illustrate simulating one season and display the data frame `RESULTS` that is returned.

```
RESULTS <- one.simulation.68(0.20)
RESULTS
```

	Team	League	Talent	Wins	Winner.Lg	Winner.WS
1	ATL	1	0.080997187	93	0	0
2	BAL	2	-0.219407302	70	0	0
3	BOS	2	0.136735454	93	0	0
4	CAL	2	-0.410259781	58	0	0
5	CHA	2	-0.142939718	72	0	0
6	CHN	1	-0.007529692	78	0	0
7	CIN	1	0.130597992	84	0	0
8	CLE	2	-0.261658757	79	0	0
9	DET	2	-0.006537641	77	0	0
10	HOU	1	-0.332069747	63	0	0
11	LAN	1	0.004409475	79	0	0
12	MIN	2	-0.180103045	81	0	0
13	NYA	2	0.105354274	86	0	0
14	NYN	1	-0.190907202	61	0	0
15	OAK	2	0.184825855	91	0	0

16	PHI	1	-0.057648893	79	0	0
17	PIT	1	0.678827397	111	1	1
18	SFN	1	-0.191052425	85	0	0
19	SLN	1	-0.230921646	77	0	0
20	WS2	2	0.451958661	103	1	0

A new function `display.standings` is written to put the season wins in a more familiar standings format. The inputs to this function are the RESULTS data frame and the league indicator.

```
display.standings <- function(RESULTS, league){
  Standings <- subset(RESULTS, League == league)[, c("Team", "Wins")]
  Standings$Losses <- 162 - Standings$Wins
  Standings[order(Standings$Wins, decreasing=TRUE), ]
}
```

This function is applied twice (once for each league) and the `cbind` function combines the two standings into a single data frame. The league champions and the World Series winner are also displayed.

```
cbind(display.standings(RESULTS, 1), display.standings(RESULTS, 2))
   Team Wins Losses Team Wins Losses
17  PIT  111     51  WS2  103     59
1   ATL   93     69  BOS   93     69
18  SFN   85     77  OAK   91     71
7   CIN   84     78  NYA   86     76
11  LAN   79     83  MIN   81     81
16  PHI   79     83  CLE   79     83
6   CHN   78     84  DET   77     85
19  SLN   77     85  CHA   72     90
10  HOU   63     99  BAL   70     92
14  NYN   61    101  CAL   58    104
with(RESULTS, as.character(Team[Winner.Lg == 1]))
[1] "PIT" "WS2"
with(RESULTS, as.character(Team[Winner.WS == 1]))
[1] "PIT"
```

In this particular simulated season, Pittsburgh (PIT) won the National League with 111 wins and Washington (WS2) won the American League with 103 wins. Pittsburgh defeated Washington in the World Series. The team with the best talent in this season was Pittsburgh (talent equal to 0.6788) and they won the World Series. In other words the "best team in baseball" was also the most successful during this simulated season. We will shortly see if the best team typically wins the World Series.

9.3.7   Simulating many seasons

One can learn about the relationship between a team's ability and its season performance by simulating many seasons of baseball. We initially set a

Many.Results data frame to NULL and use a for loop to repeat the simulation for 1000 seasons, storing the output in Many.Results.

```
Many.Results <- NULL
for(j in 1:1000)
  Many.Results <- rbind(Many.Results, one.simulation.68(0.20))
```

The data frame Many.Results contains the talent number and number of wins for $1000 \times 20 = 20,000$ teams. The smoothScatter function is used to construct a smoothed scatterplot of Talent and Wins and Figure 9.1 shows the result. (Here the plot function would have resulted in an overly cluttered scatterplot.)

```
with(Many.Results, smoothScatter(Talent, Wins))
```

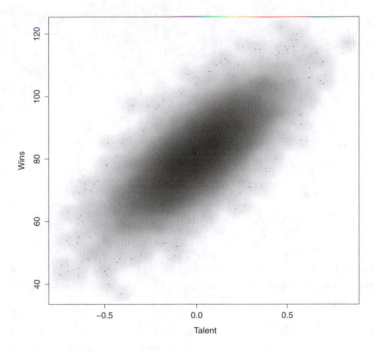

FIGURE 9.1
Smoothed scatterplot of talent and number of season wins for teams in 1000 simulated seasons.

As expected, there is a positive trend in the graph, indicating that better teams tend to win more games. But there is much vertical spread in the scatterplot which says that the relationship between talent and wins is not strong.

To reinforce the last point, suppose we focus on "average" teams that have a talent number between −0.05 and 0.05. Using the subset function, a new data frame `Results.avg` is created containing the talent and wins data for these average teams. A histogram is constructed of the season wins for these teams. (See Figure 9.2.)

```
Results.avg <- subset(Many.Results, Talent > -0.05 & Talent < 0.05)
hist(Results.avg$Wins)
```

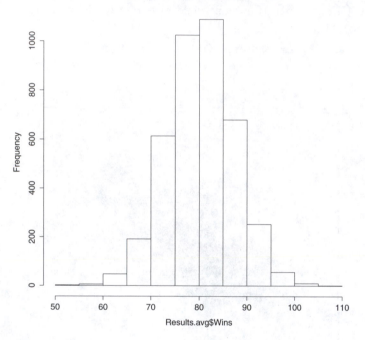

FIGURE 9.2
Histogram of the number of season wins for "average" teams in the 1000 simulated seasons.

One expects these average teams to win about 80 games. But what is surprising is the variability in the win totals – average teams can regularly have win totals between 70 and 90, and it is possible (but not likely) to have a win total close to 100.

What is the relationship between a team's talent and its post-season success? Consider first the relationship between a team's talent (variable `Talent`) and winning the league (the variable `Winner.Lg`). Since `Winner.Lg` is a binary (0 or 1) variable, a common approach for representing this relationship is a logistic model – this is a generalization of the usual regression model where the response variable is binary instead of continuous. The `glm` function with the `family=binomial` argument is used to fit a logistic model – the output is stored in the variable `fit1`. In a similar fashion, a logistic model is used to model the relationship between winning the World Series (variable `Winner.WS`) and the talent – the output is in the variable `fit2`.

```
fit1 <- glm(Winner.WS ~ Talent, data = Many.Results, family=binomial)
fit2 <- glm(Winner.Lg ~ Talent, data = Many.Results, family=binomial)
```

A logistic model has the form

$$p = \frac{\exp(a + bT)}{1 + \exp(a + bT)},$$

where T is a team's talent, (a, b) are the regression coefficients, and p is the probability of the event. In the following code, the regression coefficients of the "win pennant" logistic fit are stored in the variable `b1`. By use of the `curve` function, the fitted probability of winning the pennant is graphed as a function of the talent. A second application of `curve` is used to overlay the fitted probability of winning the World Series. The completed graph is displayed in Figure 9.3.

```
b1 <- coef(fit1)
curve(exp(b1[1] + b1[2] * x) / (1 + exp(b1[1] + b1[2] * x)),
      -0.4, 0.4, xlab="Talent", ylab="Probability", lwd=2,
      ylim=c(0, 1))
b2 <- coef(fit2)
curve(exp(b2[1] + b2[2] * x) / (1 + exp(b2[1] + b2[2] * x)),
      add=TRUE, lwd=2, lty=2)
legend(-0.2, 0.8, legend=c("Win Pennant", "Win World Series"),
                  lwd=2, lty=c(1, 2))
```

As expected, the chance of a team winning the pennant (solid line) increases as a function of the talent. An average team with $T = 0$ has only a small chance of winning the pennant; an excellent team with a talent close to 0.4 has about a 60% chance of winning the pennant. The probabilities of winning the World Series (represented by a dashed line) are substantially smaller than the chances of winning the pennant. For example, this excellent ($T = 0.4$) team has only about a 35% chance of winning the World Series. In fact, it can be demonstrated that the team winning the World Series is likely not to be the team with the best talent (largest value of T).

9.4 Further Reading

A general description of the Markov chain probability model is contained in Kemeny and Snell (1976). Pankin (1987) and Bukiet et al. (1997) illustrate the use of Markov chains to model baseball. Chapter 9 of Albert (2003) gives

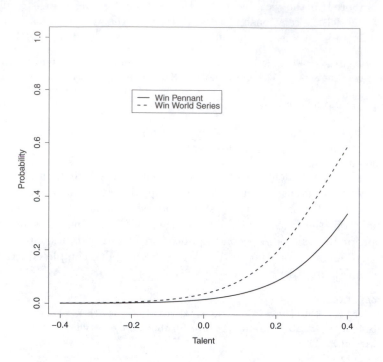

FIGURE 9.3
Probability of winning the league and the World Series for teams of different talents.

an introductory description of Markov chains and illustrates the construction and use of the transition matrix using 1987 season data. The Bradley-Terry model (Bradley and Terry, 1952) is a popular statistical model for paired comparisons. Chapter 9 of Albert and Bennett (2003) describes the application of the Bradley-Terry model for baseball team competition. The use of R in simulation is introduced in Chapter 11 of Albert and Rizzo (2012).

9.5   Exercises

1. **(A Simple Markov Chain)**

   Suppose one is interested only in the number of outs in an inning. There are four possible states in an inning (0 outs, 1 out, 2 outs, and 3 outs) and you move between these states in each plate appearance. Suppose at each PA, the chance of not increasing the number of outs is 0.3, and the probability of increasing the outs by one is 0.7. The following R code puts the transition probabilities of this Markov chain in a matrix P.

```
P <- matrix(c(.3, .7,  0,  0,
               0, .3, .7,  0,
               0,  0, .3, .7,
               0,  0,  0,  1), 4, 4, byrow=TRUE)
```

(a) If one multiplies the matrix P by itself P to obtain the matrix P2:

```
P2 <- P %*% P
```

The first row of P2 gives the probabilities of moving from 0 outs to each of the four states after two plate appearances. Compute P2. Based on this computation, find the probability of moving from 0 outs to 1 out after two plate appearances.

(b) The fundamental matrix N is computed as

```
N <- solve(diag(c(1, 1, 1)) - P[-4, -4])
```

The first row gives the average number of PAs at 0 out, 1 out, and 2 outs in an inning. Compute N and find the average number of PAs in one inning in this model.

2. **(A Simple Markov Chain, Continued)**

The following function `simulate.half.inning` will simulate the number of plate appearances in a single half-inning of the Markov chain model described in Exercise 1 where the input P is the transition probability matrix.

```
simulate.half.inning <- function(P){
  s <- 1; path <- NULL;
  while(s < 4){
    s.new <- sample(1 : 4, 1, prob=P[s, ])
    path <- c(path, s.new)
    s <- s.new
  }
  length(path)
}
```

(a) Use the `replicate` function to simulate 1000 half-innings of this Markov chain and store the lengths of these simulated innings in the vector `lengths`.

(b) Using this simulated output, find the probability that a half-inning contains exactly four plate appearances.

(c) Use the simulated output to find the average number of PAs in a half-innings. Compare your answer with the exact answer in Exercise 1, part (b).

3. (**Simulating a Half Inning**)

In Section 9.2.4, the expected number of runs as calculated for each one of the 24 possible runners-outs situations using data from the 2011 season. To see how these values can change across seasons, download play-by-play data from Retrosheet for the 1968 season, construct the probability transition matrix, simulate 10,000 half-innings from each of the 24 situations, and compute the runs expectancy matrix. Compare this 1968 runs expectancy matrix compares with the one computed using 2011 data.

4. (**Simulating the 1950 Season**)

Suppose you are interested in simulating the 1950 regular season for the American League. In this season, the team abbreviations were "PHI," "BRO," "NYG," "BSN," "STL," "CIN," "CHC," and "PIT" and each team played every other team 22 games (11 games at each park).

   (a) Using the function `make.schedule`, construct the schedule of games for this AL season.

   (b) Suppose the team talents follow a normal distribution with mean 0 and standard deviation 0.25. Using the Bradley-Terry model, assign home win probabilities for all games on the schedule.

   (c) Use the `rbinom` function to simulate the outcomes of all 616 games of the AL 1950 season.

   (d) Compute the number of season wins for all teams in your simulation.

5. (**Simulating the 1950 Season, Continued**)

   (a) Write a function to perform the simulation scheme described in Exercise 4. Have the function return the team with the largest talent and the team with the most wins. (If there is a tie for the league pennant, have the function return one of the best teams at random.)

   (b) Repeat this simulation for 1000 seasons, collecting the most talented team and the most successful team for all seasons.

   (c) Based on your simulations, what is the chance that the most talented team wins the pennant?

6. (**Simulating the World Series**)

   (a) Write a function to simulate a World Series. The input is the the probability `p` the AL team will defeat the NL team in a single game.

   (b) Suppose an AL team with talent 0.40 plays a NL team with talent 0.25. Using the Bradley-Terry model, determine the probability `p` that the AL wins a game.

   (c) Using the value of `p` determined in part (b), simulate 1000 World Series and find the probability the AL team wins the World Series.

(d) Repeat parts (b) and (c) for AL and NL teams who have the same talents.

10

Exploring Streaky Performances

CONTENTS

10.1	Introduction	237
10.2	The Great Streak	238
	10.2.1 Finding game hitting streaks	238
	10.2.2 Moving batting averages	240
10.3	Streaks in Individual At-Bats	242
	10.3.1 Streaks of hits and outs	242
	10.3.2 Moving batting averages	243
	10.3.3 Finding hitting slumps for all players	243
	10.3.4 Were Suzuki and Ibanez unusually streaky?	246
10.4	Local Patterns of Weighted On-Base Average	249
10.5	Further Reading	255
10.6	Exercises	257

10.1 Introduction

Some of the most interesting phenomena in baseball are streaky or hot/cold performances by hitters and pitchers. During particular periods in the season, a particular player will hit for a high batting average, and in other periods, the player will be in a "cold streak" and all batted balls appear to be fielded for outs. In this chapter, we'll use R to explore streaky hitting performances.

One of the great hitting accomplishments in baseball history is Joe DiMaggio's 56-game hitting streak and Section 10.2 explores DiMaggio's game-to-game hitting for the 1941 season. An R function is used to find all of DiMaggio's hitting streaks, and a moving average function is used to explore DiMaggio's batting average over short time intervals. Retrosheet play-by-play data records batters' performances in all plate appearances and we use this data in Section 10.3 to explore hitting streaks in individual at-bats. Suppose a hitter is going through an "0 for 20" hitting slump – should we be surprised? One way of answering this question is to find the longest hitting slumps for all hitters in a particular baseball season. A second way to understand the size of this hitting slump is to contrast this hitting with pattern of slumps under a random model. A method for simulating a random pattern of hits and outs is described and this method is used to assess if a particular player exhibits more streakiness in his hitting sequence than what one would expect by chance.

This discussion of streakiness focuses on patterns of hits and outs, and certainly the quality of an at-bat depends on more than just getting a hit. Section 10.4 discusses patterns of streakiness using the players' weighted on-base percentage (*wOBA*) where positive outcomes of a plate appearance are weighted by their run values. We look at players' *wOBA* over groups of five games during a season. A way to describe streaky hitting behavior is to look at the variability of the five-game *wOBA* values. Using this measure of streakiness, we identify the streaky hitters during the 2011 season.

10.2   The Great Streak

10.2.1   Finding game hitting streaks

Whenever there is a discussion of great streaky performances in baseball, one has to talk about the "Great Streak" where Joe DiMaggio hit in 56 consecutive games during the 1941 season. Many people think that this particular hitting accomplishment is one of the few baseball records that will not be broken in our lifetime. We use DiMaggio's game-to-game hitting data to motivate how we can use R to explore streaky performances.

Play-by-play hitting records are currently not available in Retrosheet for the 1941 season, but baseball-reference.com gives a game-to-game hitting log for DiMaggio for this season. By cutting and pasting the data table from the web page to a text file, we create a new file "dimaggio.1941.csv" that contains this hitting log saved in csv format. By use of the `read.csv` function, we load this file into R, creating a new data frame `joe`.

```
joe <- read.csv("dimaggio.1941.csv")
```

For each game during the season, the data frame records `AB`, the number of at-bats, and `H`, the number of hits. As a quick check that the data has been entered correctly, we compute DiMaggio's season batting average by summing the game hit totals and dividing by the total at-bats.

```
sum(joe$H) / sum(joe$AB)
[1] 0.3567468
```

The result agrees with DiMaggio's 1941 batting average of .357. (Actually, although this was a high average, it was overshadowed by Ted Williams' .406 average during the 1941 season.)

A hitting streak is commonly defined as the number of consecutive games in which a player gets at least one base hit. Suppose we're interested in computing all of DiMaggio's hitting streaks for the 1941 season. Towards this goal, using the `ifelse` function, we create a new variable `HIT` for each game that is either 1 or 0 depending on whether DiMaggio recorded at least one hit in the game.

```
joe$HIT <- ifelse(joe$H >= 1, 1, 0)
```

We display the values of `HIT` that visually shows DiMaggio's streaky hitting performance.

```
joe$HIT
  [1] 1 1 1 1 1 1 1 1 0 0 0 1 1 1 0 1 1 0 0 1 0 0 0 1 1 1 0 0 1 1 1 1
 [33] 1 1 1 1 1 1 1 1 1 1 1 1 1 1 1 1 1 1 1 1 1 1 1 1 1 1 1 1 1 1 1 1
 [65] 1 1 1 1 1 1 1 1 1 1 1 1 1 1 1 1 1 1 1 1 0 1 1 1 1 1 1 1 1 1 1 1
 [97] 1 1 1 1 1 0 0 1 1 1 1 0 0 1 1 0 0 0 1 1 1 1 0 0 0 1 1 1 1 1 1 1
[129] 0 1 0 1 1 1 1 1 0 1 1
```

We see that DiMaggio started the season with an eight-game hitting streak, then had three games with no hits, a hitting streak of three games, and so on.

Suppose we wish to compute all hitting streaks for a particular player. This is conveniently done using the following user-defined function **streaks**. The input to this function is a vector y of 0s and 1s corresponding to game results where the player was hitless (0) or received at least one hit (1). The output will be a vector containing the lengths of all hitting streaks.

```
streaks <- function(y){
  n <- length(y)
  where <- c(0, y, 0) == 0
  location.zeros <- (0 : (n+1))[where]
  streak.lengths <- diff(location.zeros) - 1
  streak.lengths[streak.lengths > 0]
}
```

In the function, we define `n` to the number of values in the vector `y`. A zero is appended to the beginning and end of `y`, and a vector `location.zeros` is defined giving the indices of the vector locations containing 0s. The lengths of all streaks of 1s are found using the function `diff` to compute the differences in the locations with 0s.

The function **streaks** is read into R and this function is applied to DiMaggio's game hit/no-hit sequence stored in the variable `joe$HIT`.

```
streaks(joe$HIT)
 [1]  8  3  2  1  3 56 16  4  2  4  7  1  5  2
```

This function picks up DiMaggio's famous 56-game hitting streak. It is interesting to note that Joe followed his 56-game streak immediately with a 16-game hitting streak.

The media is also fascinated with streaks of no-hit games. One can find DiMaggio's streaks of hitless games by creating a new variable `NO.HITS` that records 1 or 0 if the player has respectively no hits or hits, and then applying the **streaks** function to the new variable.

```
joe$NO.HIT <- 1 - joe$HIT
streaks(joe$NO.HIT)
 [1] 3 1 2 3 2 1 2 2 3 3 1 1 1
```

It is interesting that the length of the longest streak of no-hit games was only three for DiMaggio's 1941 season.

10.2.2   Moving batting averages

An alternative way of looking at streaky hitting performances uses batting averages computed over short time intervals. One may be interested in exploring DiMaggio's batting average in this manner. He must have been a hot hitter during his 56-game hitting streak, and perhaps DiMaggio was somewhat cold in other periods during the season.

In general, suppose we are interested in computing a player's batting average over a width or window of 10 games. We want to compute the batting average over games 1 to 10, over games 2 to 11, over games 3 to 12, and so on. These batting averages would be the sum of hits divided by the sum of at-bats over the 10-game periods. These short-term batting averages are commonly called *moving averages*.

The function `moving.average` below computes these moving averages. The arguments to the function are the vector of hits `H`, the vector of at-bats `AB`, and the window of games `width`.

```
moving.average <- function(H, AB, width){
  require(TTR)
  game <- 1 : length(H)
  P <- data.frame(Game=SMA(game, width),
        Average=SMA(H, width) / SMA(AB, width))
  P[complete.cases(P), ]
}
```

A vector `game` is defined to be the integers 1 through n, where n is the number of games. The function `SMA` from the package `TTR` computes the simple moving averages of `game`, `H` and `AB`, and those quantities are used to compute the moving batting average `AVG`. The output of the function is a data frame with two variables `Game` and `Average`. The variable `Game` gives the game number value in the middle of the window, and `Average` is the corresponding batting average over the game window.

After the function `moving.average` is read into R, it is easy to compute DiMaggio's batting average over short time intervals. Suppose we consider a window of 10 games. In the following code, we use `moving.average` to compute the moving batting averages and the output is passed to `plot` to construct a line graph of these averages (see Figure 10.1). A horizontal line is added at DiMaggio's season batting average so one can easily see when Joe was relatively hot and cold during the season. To relate this display with DiMaggio's hitting streaks, we use the `rug` function to display the games where Joe had at least one hit on the horizontal axis.

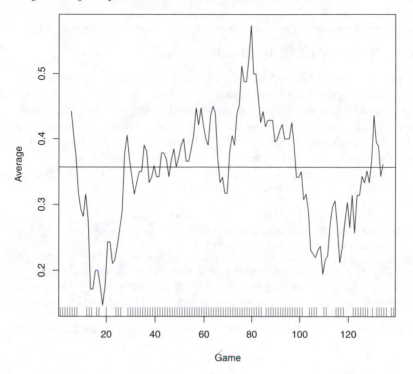

FIGURE 10.1
Moving average plot of DiMaggio's batting average for the 1941 season using a window of 10 games. The horizontal line shows DiMaggio's season batting average. The games where DiMaggio had at least one base hit are displayed on the horizontal axis.

```
plot(moving.average(joe$H, joe$AB, 10), type="l")
abline(h=sum(joe$H) / sum(joe$AB))
game.hits <- (1:nrow(joe))[as.logical(joe$HIT)]
rug(game.hits)
```

This figure dramatically shows that DiMaggio's hitting performance climbed steadily during his 56-game hitting streak and he actually had a short-term 10-game batting average over .500 during the streak. DiMaggio had a noticeable hitting slump in the second half of the season and he hit bottom about game 110. In practice, the appearance of this graph may depend on the choice of time interval (argument `width` in the function `moving.average`) and one should experiment with several `width` choices to get a better understanding of a hitter's short-term batting performance.

10.3   Streaks in Individual At-Bats

The previous section considered hitting streaks at a game-to-game level. Since
records of individual plate appearances are available in the Retrosheet play-
by-play files, it is straightforward to explore hitting streaks at this finer level.
Ichiro Suzuki is one of the most exciting hitters in baseball, especially for his
ability to hit singles, many of the infield variety. The streakiness patterns in
Suzuki's play-by-play hitting data is explored for the 2011 season.

We begin by reading the Retrosheet play-by-play file for the 2011 season,
storing the file in the data frame data2011.

```
data2011 <- read.csv("all2011.csv", header=FALSE)
fields <- read.csv("fields.csv")
names(data2011) <- fields[, "Header"]
```

The subset function is used to define a new data frame ichiro.AB; records
are chosen where the batting id is "suzui001" (Suzuki's code id) and the at-bat
flag is TRUE. (In this exploration, only Suzuki's official at-bats are considered.)

```
ichiro.AB <- subset(data2011, BAT_ID == "suzui001" & AB_FL == TRUE)
```

10.3.1   Streaks of hits and outs

We record at each at-bat if a hit occurred. There is a current variable H_FL
in the data2011 data frame recording the number of bases for a hit. Using
the ifelse function, a new variable HIT is defined that is 1 if a hit occurs
and 0 otherwise. To make sure that these at-bats are correctly ordered in time
during the season, a variable DATE (extracted from the GAME_ID variable using
the substr function) is defined, and the order function sorts the data frame
ichiro.AB by date.

```
ichiro.AB$HIT <- ifelse(ichiro.AB$H_FL > 0, 1, 0)
ichiro.AB$DATE <- substr(ichiro.AB$GAME_ID, 4, 12)
ichiro.AB <- ichiro.AB[order(ichiro.AB$DATE), ]
```

From the variable HIT, the lengths of all hitting streaks are identified,
where a streak refers to a sequence of consecutive base hits. Using the streaks
function defined in Section 10.2, the streak lengths are obtained for Suzuki in
the 2011 season.

```
streaks(ichiro.AB$HIT)
  [1] 1 1 1 2 1 1 1 1 1 1 1 1 2 1 1 1 1 2 2 1 1 1 1 4 1 1 1 1 2 3 1 1
 [31] 1 1 1 1 1 2 2 1 1 1 1 1 2 2 1 2 2 1 1 1 2 1 2 1 1 2 1 3 1 3
 [61] 1 1 1 1 1 1 1 1 1 2 1 1 2 1 2 1 1 1 3 1 1 1 2 1 4 1 1 2 3 1
 [91] 1 1 1 1 1 2 1 2 1 1 1 2 1 2 1 1 1 1 1 1 1 1 1 1 1 2 1 1 1
[121] 1 1 4 1 1 1 1 2 2 1 1 3 1 1 1 2
```

As expected, most of the hitting streaks lengths are 1, although several times Suzuki had four consecutive hits.

It may be more interesting to explore the lengths of the gaps between hits. By the operation `1 - HIT`, the roles of 0 and 1 are reversed in the sequence, and the function `streaks` is applied to find the lengths of all of the gaps between hits that are 1 or larger.

```
streaks(1 - ichiro.AB$HIT)
  [1]  1  2  2  4  8  1  3  6  2  3  6  1  2  8  4  1  1  1  5  2
 [21]  1  1  4  4  1  1  4  1  5  6  2  3  1  2  4  4  5  4  3  2
 [41] 10  2  2 12 11  5 16  2  2 15  2  1  1  2  1  1  2  2  4  2
 [61]  3  3  9  1  3  2  1  3 10  2  1  1 10  3 15  6  4  3  1  2
 [81]  3  3  3  5  4 11  5  1  2  4  3  5  1  3  4  5 10  3  3  3
[101]  1  1  3  2  1  2  3  2  2  2  1  1  1  4  6  1  5  3  1  8
[121]  4  1  1  7  1  7  6  5  1  2  6  5  6  2  3  6
```

This output is more interesting; a frequency table of this output is constructed using the `table` function.

```
table(streaks(1 - ichiro.AB$HIT))

 1  2  3  4  5  6  7  8  9 10 11 12 15 16
35 28 22 15 11  9  2  3  1  4  2  1  2  1
```

It is seen that Suzuki had a streak of 12 outs once, a streak of 15 outs twice, and a streak of 16 outs once.

10.3.2 Moving batting averages

Another way to view Suzuki's streaky batting performance is to consider his batting average over short time intervals, analogous to what we did for DiMaggio for his game-to-game hitting data. Using the `moving.average` function, a moving average plot is constructed of Ichiro's batting average using a window of 30 at-bats (see Figure 10.2). Using the `rug` function, we display the at-bats where Ichiro had hits. The long streaks of outs are visible as gaps in the rug plot. During the middle of the season, Ichiro had a 30 at-bat batting average exceeding 0.500, while during other periods, his 30 at-bat average was as low as 0.100.

```
ichiro.AB$AB <- 1
plot(moving.average(ichiro.AB$HIT, ichiro.AB$AB, 30), type="l", xlab="AB")
abline(h=mean(ichiro.AB$HIT))
rug((1:nrow(ichiro))[ichiro.AB$HIT == TRUE])
```

10.3.3 Finding hitting slumps for all players

In our exploration of Suzuki's batting performance, we saw that he had a "0 for 16" hitting performance during the season. Should we be surprised by

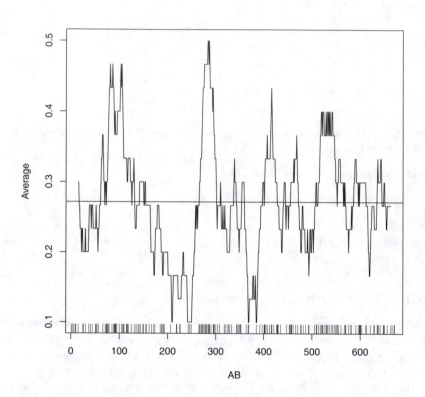

FIGURE 10.2
Moving average plot of Ichiro Suzuki's batting average for the 2011 season using a window of 30 at-bats. The horizontal line shows Suzuki's season batting average. The at-bats where Suzuki had at least one base hit are shown on the horizontal axis.

a hitting slump of length 16? Let's compare Suzuki's long slump with the longest slumps for all regular players during the 2011 season.

First a new function `longest.ofer` is written that computes the length of the longest hitting slump for a given batter. (A "ofer" is a slang word for a hitless streak in baseball.) The inputs to this function are the batter id code `batter` and the batting data frame `data`. The output of the function is the length of the longest slump.

```
longest.ofer <- function(batter, data){
  d.AB <- subset(data, BAT_ID == batter & AB_FL == TRUE)
  d.AB$HIT <- ifelse(d.AB$H_FL > 0, 1, 0)
  d.AB$DATE <- substr(d.AB$GAME_ID, 4, 12)
  d.AB <- d.AB[order(d.AB$DATE), ]
  source("streaks.R")
  max(streaks(1 - d.AB$HIT))
}
```

After reading this function into R, we confirm that it works by finding the longest hitting slump for Suzuki.

```
longest.ofer("suzui001", data2011)
[1] 16
```

Suppose we want to compute the length of the longest hitting slump for all players in this season with at least 400 at-bats. Using the `aggregate` function, the number of at-bats is computed for all players, and `players.400` is defined to be the vector of the id codes of all players with 400 or more at-bats. By use of the `sapply` function together with the `longest.ofer` function, the length of the longest slump is computed for all regular hitters. Using the `data.frame` function, the vector S is converted to a data frame, adding the variable `Player`. (The `rownames(S) <- NULL` command is used to remove the player ids from the row names.)

```
A <- aggregate(data2011$AB_FL, list(Player=data2011$BAT_ID), sum)
players.400 <- A$Player[A$x >= 400]
S <- sapply(players.400, longest.ofer, data2011)
S <- data.frame(Player=names(S), Streak=S)
rownames(S) <- NULL
```

To decipher the player ids, it is helpful to merge the data frame of the longest hitting slumps S with the player roster information contained in the Retrosheet file "roster2011.csv." This roster file is read into R, saving it in the data frame `roster2011`. Then the `merge` function is applied, merging data frames S and `roster2011`, matching on the variables `Player` (in S) and `Player.ID` (in `roster2011`).

```
roster2011 <- read.csv("roster2011.csv")
S1 <- merge(S, roster2011, by.x="Player", by.y="Player.ID")
```

The slump lengths are ordered in decreasing order using the function `order` with the `decreasing=TRUE` argument and the rows of the data frame S1 are reordered using this ordering. The top six slump lengths are displayed by the head function.

```
S.ordered <- S1[order(S1$Streak, decreasing=TRUE), ]
head(S.ordered)
        Player Streak    X Last.Name First.Name Bats Pitches Team V7
80   ibanr001     35  941    Ibanez       Raul    L       R  PHI OF
6    aybae001     30    3     Aybar      Erick    B       R  ANA SS
113  mcgec001     27  726   McGehee      Casey    R       R  MIL 3B
122  olivm001     27 1095     Olivo     Miguel    R       R  SEA  C
152  ruizc001     26  958      Ruiz     Carlos    R       R  PHI  C
48   ellim001     25  896     Ellis       Mark    R       R  OAK SS
```

The six longest hitting slumps during the 2011 season were by Raul Ibanez (35), Erick Aybar (30), Casey McGehee (27), Miguel Olivo (27), Carlos Ruiz (26), and Mark Ellis (25). Relative to these long hitting slumps, Suzuki's hitting slump of 16 at-bats looks short.

10.3.4   Were Suzuki and Ibanez unusually streaky?

In the previous section, patterns of streakiness of hit/out data were compared for all players in the 2011 season. An alternative way to look at the streakiness of a player is to contrast his streaky pattern of hitting with streaky patterns under a "random" model.

To illustrate this method, consider a hypothetical player who bats 13 times with the outcomes

$$0, 1, 0, 0, 1, 1, 0, 0, 0, 0, 1, 1, 1.$$

A measure of streakiness is defined based on this sequence of hits and outs. One good measure of streakiness or clumpiness in the sequence is the sum of squares of the gaps between successive hits. In this example, the gaps between hits are 1, 2, and 4, and the sum of squares of the gaps is $S = 1^2 + 2^2 + 4^2 = 21$.

Is the value of streakiness statistic $S = 21$ large enough to conclude that this player's pattern of hitting is non-random? This question is answered by a simple simulation experiment. If the player sequence of hit/out outcomes is truly random, then all possible arrangements of the sequence of 6 hits and 7 outs are equally likely. We randomly arrange the sequence 0, 1, 0, 0, 1, 1, 0, 0, 0, 0, 1, 1, 1, find the gaps, and compute the streakiness measure S. This randomization procedure is repeated many times, collecting, say, 1000 values of the streakiness measure S. A histogram of the values of S is constructed – this histogram represents the distribution of S under a random model. If the observed value of $S = 21$ is in the middle of the histogram, then the player's pattern of streakiness is consistent with a random model. On the other hand, if the value $S = 21$ is in the right tail of this histogram, then the observed streaky pattern is not consistent with "random" streakiness and there is evidence that the player's pattern of hits and outs is non-random.

This method is first illustrated for Ichiro Suzuki's 2011 hitting data. As before, Suzuki's hitting data is read into the data frame `ichiro.AB` – the variable `HIT` is a sequence of 0s and 1s, where 0 corresponds to an out and 1 corresponds to a hit.

```
ichiro.AB <- subset(data2011, BAT_ID == "suzui001" & AB_FL == TRUE)
ichiro.AB$HIT <- ifelse(ichiro.AB$H_FL > 0, 1, 0)
ichiro.AB$DATE <- substr(ichiro.AB$GAME_ID, 4, 12)
ichiro.AB <- ichiro.AB[order(ichiro.AB$DATE), ]
```

The clumpiness or streakiness is measured by the sum of squares of all gaps between hits. The function `streaks` is used to find the gaps which are stored in the vector `st`. Each of the gap values is squared and the `sum` function computes the sum.

```
source("streaks.R")
st <- streaks(1 - ichiro.AB$HIT)
sum(st ^ 2)
[1] 3047
```

The value of Suzuki's streakiness statistic is $S = 3047$.

Next, a function `random.mix` is written to perform one iteration of the simulation experiment where the input `y` is a vector of 0s and 1s. The `sample` function finds a random arrangement of `y`, the `streaks` function finds the gaps between hits, and the function `clump.stat` finds the sum of squares of the gaps.

```
random.mix <- function(y){
  source("streaks.R")
  clump.stat <- function(sp) sum(sp ^ 2)
  mixed <- sample(y)
  clump.stat(streaks(1 - mixed))
}
```

By use of the `replicate` function, this simulation experiment is repeated 1000 times, storing the values of the streakiness statistic in the vector `ST`.

```
ST <- replicate(1000, random.mix(ichiro.AB$HIT))
```

A histogram of the values of `ST` is constructed (using the `truehist` function in the package `MASS`) and the `abline` function is used to overlay the clumpiness value (3047) for Suzuki. (See Figure 10.3.)

```
library(MASS)
truehist(ST)
abline(v=3047, lwd=3)
text(3250, 0.0016, "Suzuki", cex=1.5)
```

Since the value of 3047 is in the middle of the histogram distribution, the streakiness pattern in Suzuki's hitting is consistent with a random model. There is insufficient evidence that Suzuki is truly streaky.

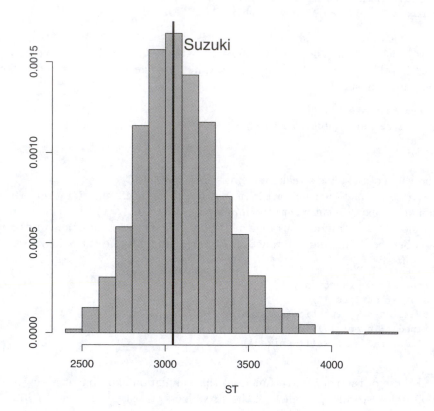

FIGURE 10.3
Histogram of one thousand values of the clumpiness statistic assuming all
arrangements of hits and outs for Suzuki are equally likely. The observed
value of the clumpiness statistic for Suzuki is shown using a vertical line.

This method can be used to check if the streaky patterns of any hitter are non-random. A new function `clump.test` is constructed using the R code previous discussed. The input is the player id code `playerid` and the season batting data frame `data`. One thousand values of the clumpiness measure are computed by 1000 replications of the simulation procedure. A histogram of the clumpiness measures is constructed and the observed clumpiness statistic is shown as a vertical line.

```
clump.test <- function(playerid, data){
  player.AB <- subset(data, BAT_ID == playerid & AB_FL == TRUE)
  player.AB$HIT <- ifelse(player.AB$H_FL > 0, 1, 0)
  player.AB$DATE <- substr(player.AB$GAME_ID, 4, 12)
  player.AB <- player.AB[order(player.AB$DATE), ]
  ST <- replicate(1000, random.mix(player.AB$HIT))
  truehist(ST, xlab="Clumpiness Statistic")
  stat <- sum((streaks(1 - player.AB$HIT)) ^ 2)
  abline(v=stat, lwd=3)
  text(stat * 1.05, 0.0010, "OBSERVED", cex=1.2)}
```

To investigate the non-randomness of Raul Ibanez's sequence of hit/out data, the function `clump.test` is run using Ibanez's player id cole "ibanr001" and the resulting histogram display is shown in Figure 10.4.

```
clump.test("ibanr001", data2011)
```

Note that Ibanez's clumpiness measure is in the right tail of this distribution, indicating that Ibanez clearly displayed more streakiness than one would expect by chance.

10.4 Local Patterns of Weighted On-Base Average

In our discussion of hitting slumps and streaks, the focus is on either getting a hit or an out in an official at-bat. A successful plate appearance goes beyond simply getting a hit or an out, and we would like to explore patterns of slumps and streaks using a better measure of hitting performance.

One popular measure of hitting performance, the weighted on-base percentage or $wOBA$, is based on giving weights to each positive outcome of a plate appearance where the weights are proportional to the run values of the outcomes. This measure is obtained by summing these weights and dividing the sum by the number of plate appearances.

In one definition, the non-intentional walks ($NIBB$), hit-by-pitches (HBP), singles ($1B$), reached base by error ($RBOE$), doubles ($2B$), triples ($3B$), and home runs (HR) are given the respective weights 0.72, 0.75, 0.90, 0.92, 1.24, 1.56, and 1.95, and $wOBA$ is given by the sum of the weights

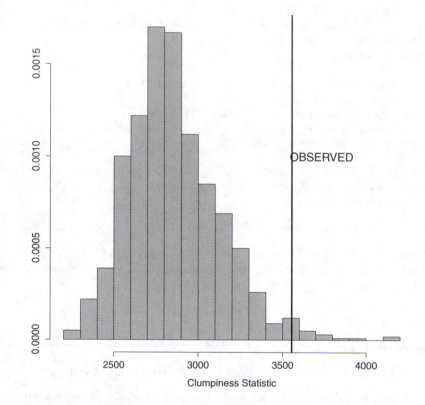

FIGURE 10.4
Histogram of one thousand values of the clumpiness statistic assuming all
arrangements of hits and outs for Raul Ibanez are equally likely. The observed
value of the clumpiness statistic for Ibanez is shown using a vertical line.

divided by the plate appearances (PA). A new function `sum.woba` is written that computes the sum of plate appearances and sum of $wOBA$ weights for a vector `event` of codes of plate appearance events. This function uses the same codes used in the Retrosheet variable "EVENT_CD". (The code 14 corresponds to a walk, the code 16 corresponds to a hit-by-pitch, and so on.)

```
sum.woba <- function(event){
  s <-   0.72 * (event == 14) + 0.75 * (event == 16) +
         0.90 * (event == 20) + 0.92 * (event == 18) +
         1.24 * (event == 21) + 1.56 * (event == 22) +
         1.95 * (event == 23)
  c(sum(s), length(s))
}
```

Since we wish to focus our exploration to plate appearances, the `subset` function is used to create a new data frame `data2011a` containing only of batting events (`BAT_EVENT_FL == TRUE`). The `aggregate` function is used together with the `sum.woba` function to find the sum of weights and number of PAs for each player for each game in the 2011 season. Looking at a few lines of the new data frame `S`, we see Bobby Abreu (abbreviation "abreb001") in the first game had a total weight of 1.44 in 4 plate appearances – the value of $wOBA$ for this game would be $1.44/4 = 0.360$.

```
data2011a <- subset(data2011, BAT_EVENT_FL == TRUE)
S <- with(data2011a,
aggregate(EVENT_CD, list(Player=BAT_ID, Game=GAME_ID), sum.woba))
head(S)
      Player         Game  x.1  x.2
1 abreb001 ANA201104080 1.44 4.00
2 bautj002 ANA201104080 0.90 4.00
3 bourp001 ANA201104080 0.00 4.00
4 calla001 ANA201104080 0.72 4.00
5 congh001 ANA201104080 0.90 4.00
6 davir003 ANA201104080 0.00 4.00
```

Say we are interested in looking at a player's $wOBA$ over groups of five games – games 1-5, games 6-10, games 11-15, and so on. A player's batting performance for all games in a season is represented as a matrix, where rows correspond to games, the first column corresponds to the $wOBA$ weights for all games, and the second column corresponds to the game PAs. The new function `regroup` collapses a player's batting performance matrix into groups of size `g`, where a particular row will correspond to the sum of weights and sum of PAs in a particular group. (In our exploration, groups of size `g` = 5 will be used.)

```
regroup <- function(d, g){
  n <- dim(d)[1]
  n.g <- floor(n / g); n.r <- n - n.g * g
  gp <- rep(1 : n.g, each=g, length.out=n)
```

```
if(n.r > 0) {
  i <- (n.g * g + 1) : n
  gp[i] <- rep(n.g, length(i))
}
aggregate(d, list(gp), sum)[, -1]
}
```

To illustrate this grouping operation, the game-to-game hitting data for Suzuki is collected in the data frame suzuki. As before, to make sure the data is chronologically ordered, a new variable Date is created and the rows are ordered by increasing values of Date. Showing the first few rows of suzuki, we see that the variable x (labeled by x.1 and x.2) is the matrix of hitting data. The regroup function is applied on the matrix suzuki$x. The output is a matrix with two columns: V1 contains the sum of weights for each group of five games and V2 is the number of PAs in each group. (Not all rows of the matrix are shown.)

```
suzuki <- subset(S, Player == "suzui001")
suzuki$Date <- with(suzuki, substr(Game, 4, 12))
suzuki <- suzuki[order(suzuki$Date), ]
head(suzuki)
           Player        Game   x.1  x.2        Date
32293 suzui001 OAK201104010 3.44 5.00 201104010
32314 suzui001 OAK201104020 1.80 5.00 201104020
32335 suzui001 OAK201104030 1.62 4.00 201104030
46061 suzui001 TEX201104040 1.24 5.00 201104040
46079 suzui001 TEX201104050 0.92 4.00 201104050
46098 suzui001 TEX201104060 1.44 4.00 201104060
regroup(suzuki$x, 5)
     V1 V2
1   9.02 23
2   6.66 23
3   5.18 23
4   7.36 22
...
30  4.32 23
31  8.56 23
32  5.22 25
```

The process of finding the five-game hitting data has been illustrated for Suzuki. When we look at the sequence of five-game wOBAs for an arbitrary player, the wOBAs for a consistent player will have small variation, and the wOBA values for a streaky player will have high variability. A common measure of variability is the standard deviation, the average size of the deviations from the mean.

A new function is written to compute the mean and standard deviation of the group wOBAs for a given player. This function get.streak.data performs this operation for a given player with id code playerid, the weight matrix for all plays S, and a grouping of g games (by default g = 5). The output is a vector with the sum of plate appearances N, the mean of the group wOBAs Mean and the standard deviation of the group wOBAs SD.

```
get.streak.data <- function(playerid, S, g=5){
  S.player <- subset(S, Player == playerid)
  S.player$Date <- with(S.player, substr(Game, 4, 12))
  S.player <- S.player[order(S.player$Date), ]
  S.player.gp <- regroup(S.player$x, g)
  s.woba.avg <- with(S.player.gp, V1 / V2)
  c(N=sum(S.player.gp$V2),
          Mean=mean(s.woba.avg), SD=sd(s.woba.avg))
}
```

To illustrate the use of this function, we apply it to Suzuki's hitting data.

```
get.streak.data("suzui001", S, 5)
            N          Mean            SD
721.00000000     0.28498259    0.09212598
```

Suzuki had 721 plate appearances, the mean of his five-game wOBAs was 0.285 and the standard deviation of his five-game wOBAs was 0.092.

The function get.streak.data is applied to all players in the 2011 season. The vector player.list is defined to be the vector of all unique player ids and the sapply function is used to apply get.streak.data to all players in player.list. We want to focus on "regular" players and the subset function is used to collect the streakiness data only for players where the number of plate appearances (variable N) is 500 or greater.

```
player.list <- unique(S$Player)
Results <- data.frame(Player=player.list,
                 t(sapply(unique(S$Player),
                 get.streak.data, S)))
Results.500 <- subset(Results, N >= 500)
```

A scatterplot is constructed of the mean and standard deviation of wOBAs of all regular players in Figure 10.5. Using the identify function, we locate the points corresponding to the largest and smallest standard deviations and place the last names of these players on the graph (using information from the Master data frame).

```
Master <- read.csv("Master.csv")
with(Results.500, plot(Mean, SD))
ids <- with(Results.500, identify(Mean, SD, n=2, Player, plot=FALSE))
pids <- as.character(Results.500[ids, "Player"])
with(Results.500, text(Mean[ids], SD[ids],
     subset(Master, retroID %in% pids)$nameLast, pos=2))
```

The streakiest hitter during the 2011 season using this standard deviation measure was Justin Upton. Likewise, the most consistent player, Ichiro Suzuki, is identified as the one with the smallest standard deviation of the period

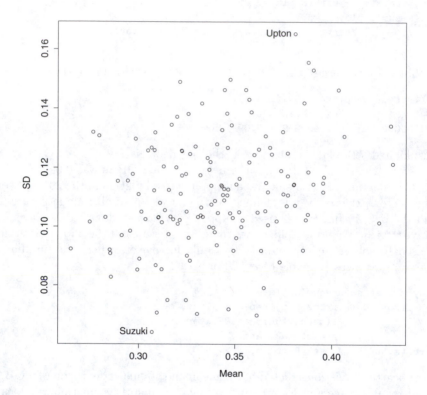

FIGURE 10.5
Scatterplot of mean and standard deviation of five-game wOBAs for all players
in the 2011 with at least 500 PA. Two players are identified, Suzuki and Upton,
who had small and large standard deviations, respectively.

wOBAs. These two players can be compared graphically by plotting their five-game wOBA values against the period number.

A new function `get.streak.data2` is created to compute the vector of five-game wOBA values for a particular player. This function is a simple modification of the function `get.streak.data` where the last line is replaced by the line:

```
data.frame(Period=1:length(s.woba.avg), wOBA=s.woba.avg)
```

The function returns the period number `Period` and the corresponding weighted on-base percentage `wOBA`.

Using this new function, a data frame `d1` is created with Suzuki's data and similar data frame `d2` is created for Upton's data and the two data frames are merged by use of the `rbind` function. The graphics functions `ggplot`, `geom_line`, and `facet_grid` in the `ggplot2` package are used to create the line graphs. One nice feature of `ggplot2` graphics is that it automatically uses the same vertical scale for the two panels and shows the player names on the right of the graph.

```
d1 <- get.streak.data2("suzuk001", S)
d2 <- get.streak.data2("uptoj001", S)
d <- rbind(data.frame(Player="Suzuki", d1),
           data.frame(Player="Upton", d2))
library(ggplot2)
ggplot(d, aes(Period, wOBA)) +
geom_line(size=1) + facet_grid(Player ~ . )
```

Note that, as expected, Suzuki and Upton have dramatically different patterns of five-game wOBAs. Most of Suzuki's five-game wOBAs fall between 0.200 and 0.400. In contrast, Upton had a change in wOBA of 0.000 to 0.740 in two periods; he was a remarkably streaky hitter during the 2011 season.

10.5 Further Reading

There is much interest in streaky performances of baseball players in the literature. Gould (1989), Berry (1990), and Seidel (2002) discuss the significance of DiMaggio's hitting streak in the 1941 season. Albert and Bennett (2003), Chapter 5, describe the difference between observed streakiness and true streakiness and give an overview of different ways of detecting streakiness of hitters. Albert (2008) and McCotter (2009) discuss the use of randomization methods to detect if there is more streakiness in hitting data than one would expect by chance.

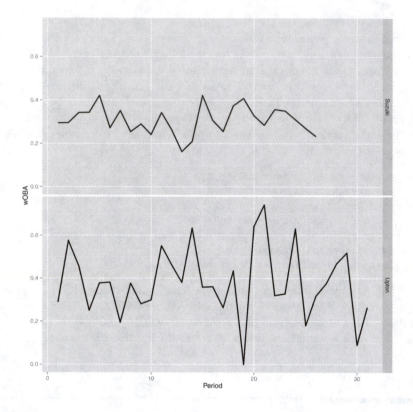

FIGURE 10.6
Line plots of five-game wOBAs against period number for Ichiro Suzuki and
Justin Upton for the 2011 season. Suzuki had a very consistent pattern of
wOBAs, while Upton's pattern of wOBAs is very volatile.

10.6 Exercises

1. **(Ted Williams)**

   The data file "williams.1941.csv" contains Ted Williams game-to-game hitting data for the 1941 season. This season was notable in that Williams had a season batting average of .406 (the most recent season batting average exceeding .400). Read this dataset into R.

   (a) Using the R function `streaks`, find the lengths of all of Williams' hitting streaks during this season. Compare the lengths of this hitting streaks with those of Joe DiMaggio during this same season.

   (b) Use the function `streaks` to find the lengths of all hitless streaks of Williams during the 1941 season. Compare these lengths with those of DiMaggio during the 1941 season.

2. **(Ted Williams, Continued)**

   (a) Use the R function `moving.average` to find the moving batting averages of Williams for the 1941 season using a window of 5 games. Graph these moving averages and describe any hot and cold patterns in Williams hitting during this season.

   (b) Compute and graph moving batting averages of Williams using several alternative choices for the window of games.

3. **(Streakiness of the 2008 Lance Berkman)**

   Lance Berkman had a remarkable hot period of hitting during the 2008 season.

   (a) Download the Retrosheet play-by-play data for the 2008 season, and extract the hitting data for Berkman.

   (b) Using the function `streaks`, find the lengths of all hitting streaks of Berkman. What was the length of his longest streak of consecutive hits?

   (c) Use the `streaks` function to find the lengths of all streaks of consecutive outs. What was Berkman's longest "oh-for" during this season?

   (d) Construct a moving batting average plot using a window of 20 at-bats. Comment on the patterns in this graph – was there a period when Berkman was unusually hot?

4. **(Streakiness of the 2008 Lance Berkman, Continued)**

   (a) Use the method described in Section 10.3.4 to see if Berkman's streaky patterns of hits and outs are consistent with patterns from a random model.

(b) The method of Section 10.3.4 used the sum of squares of the gaps as a measure of streakiness. Suppose one uses the longest streak of consecutive outs as an alternative measure. Rerun the method with this new measure and see if Berkman's longest streak of outs is consistent with the random model.

5. (**Streakiness of All Players During the 2008 Season**)

   (a) Using the 2008 Retrosheet play-by-play data, extract the hitting data for all players with at least 400 at-bats.

   (b) For each player, find the length of the longest streak of consecutive outs. Find the hitters with the longest streaks and the hitters with shortest streaks. How does Berkman's longest "oh-for" compare in the group of longest streaks?

6. (**Streakiness of All Players During the 2008 Season, Continued**)

   (a) For each player and each game during the 2008 season, compute the sum of $wOBA$ weights and the number of plate appearances PA. (See Section 10.4.)

   (b) For each player with at least 500 PA, compute the $wOBA$ over groups of five games (games 1-5, games 6-10, etc.) For each player, find the standard deviation of these five-game $wOBA$, and find the ten most streaky players using this measure.

7. (**The Great Streak**)

   The Retrosheet website recently added play-by-play data for the 1941 season when Joe DiMaggio achieved his 56-game hitting streak.

   (a) Download the 1941 play by play data from the Retrosheet website.

   (b) Confirm that DiMaggio had three "0 for 12" streaks during the 1941 season.

   (c) Use the method described in Section 10.3.4 to see if DiMaggio's streaky patterns of hits and outs in individual at-bats are consistent with patterns from a random model.

   (d) DiMaggio is perceived to be very streaky due to his game-to-game hitting accomplishment during the 1941 season. Based on your work, is DiMaggio's pattern of hitting also very streaky on individual at-bats?

11

Learning About Park Effects by Database Management Tools

CONTENTS

11.1	Introduction	259
11.2	Installing MySQL and Creating a Database	260
11.3	Connecting R to MySQL	262
	11.3.1 Connecting using package `RMySQL`	262
	11.3.2 Connecting using Package `RODBC`	263
11.4	Filling a MySQL Game Log Database from R	264
	11.4.1 From Retrosheet to R	265
	11.4.2 From R to MySQL	265
11.5	Querying Data from R	268
	11.5.1 Introduction	268
	11.5.2 Coors Field and run scoring	271
11.6	Baseball Data as MySQL Dumps	273
	11.6.1 Lahman's database	273
	11.6.2 Retrosheet database	274
	11.6.3 PITCHf/x database	274
11.7	Calculating Basic Park Factors	275
	11.7.1 Loading the data in R	275
	11.7.2 Home run park factor	276
	11.7.3 Assumptions of the proposed approach	277
	11.7.4 Applying park factors	278
11.8	Further Reading	279
11.9	Exercises	279

11.1 Introduction

In this book, analyses were performed entirely from baseball datasets loaded into R. That was possible because we were dealing with datasets with a relatively small number of rows. However, when one wants to work on multiple seasons of play-by-play (or pitch-by-pitch) data, it become more difficult to manage all of the data inside R.[1] While Retrosheet gamelogs consist of approximately 250,000 records, there are approximately 10 million Retrosheet play-by-play events, and MLBAM provides data on roughly 800,000 pitches per year for MLB games.

A solution to this "big data" problem is to store the data using a Database Management System (DBMS) and, by communicating the system with R, access only the data needed for the particular analysis. In this chapter some guidance is provided on this task. Our choice for the DBMS is MySQL, likely the most popular open-source DBMS. However, readers familiar with other software can find similar solutions for their DBMS of choice.

The use of MySQL and the R interface are used to gain some understanding of park effects in baseball. Unlike most of the other team sports, baseball ballparks vary greatly in shape and dimensions. The left-field wall in Fenway Park, home of the Boston Red Sox, is listed at 310 feet from home plate, while the left-field fence in Wrigley Field (home of the Chicago Cubs) is 355 feet away. The left-field wall in Boston, commonly known as The Green Monster, is 37 feet high, while the left-field fence Dodger Stadium in Los Angeles is only four feet high. Such differences in ballpark shapes and dimensions and the prevalent weather conditions have a profound effect on the game and the associated player measures of performance.

We first show how to obtain and set up MySQL, and then illustrate connecting R to a MySQL database for the purpose of retrieving data in R and appending data to MySQL tables. This interface is used to present evidence of the effect of Coors Field (home of the Colorado Rockies) on run scoring. The reader is directed to online resources providing baseball data (of the seasonal to pitch-by-pitch type) ready for import into MySQL. The chapter is concluded by providing the readers with a basic approach for calculating park factors and using these factors to make suitable adjustments to players' stats.

11.2 Installing MySQL and Creating a Database

Since this is a book focusing on R, we emphasize the use of MySQL together with R. A user can install MySQL on its own from `www.mysql.com/downloads`. Alternatively, one can obtain this database software by the installation of XAMPP, a distribution containing an Apache Web server, plus MySQL, PHP, and Perl, freely available for Linux, Windows, Mac OS X, and Solaris. On the XAMPP Web page[2] one finds information on how to install the distribution for the desired platform. Below we show examples related to

---

[1] R by default reads data into memory (RAM) thus imposing limits on the size of datasets it can read.

[2] `www.apachefriends.org/en/xampp.html`

the Windows distribution, but readers using other operating systems should find a similar installation procedure.

If one has not modified the defaults during the installation, XAMPP can be launched from the Programs option under the Start Menu. In the box that appears (Figure 11.1) one clicks on the *Start* button. beside *Apache* and *MySQL* to start those services. By clicking on *Admin...* on the right side of *MySQL* the default browser is opened and one is taken to the `localhost/phpmyadmin` page (Figure 11.2).

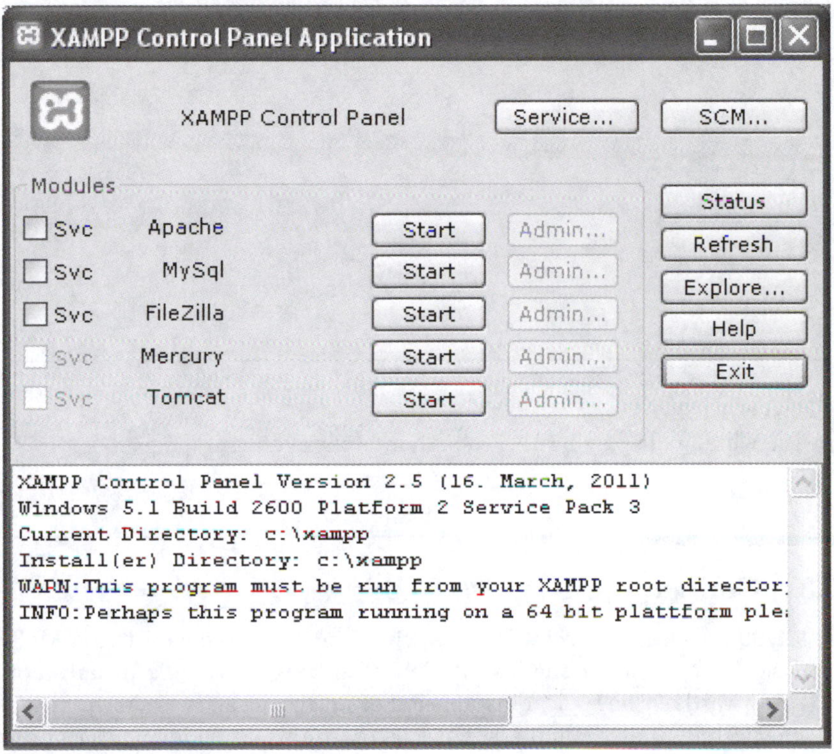

FIGURE 11.1
XAMPP Control Panel Application.

Let's now create a new database named `RBaseball`. First, click on the *Databases* tab, then type `RBaseball` on the text box under *Create new database* and click the *Create* button. After a few seconds one should see the `RBaseball` database listed on the left frame.[3] If one clicks on the newly created database, a message will appear saying there are no tables in the database.

---

[3]If that is not the case, simply clicking the *Reload navigation frame* green arrow will make it appear.

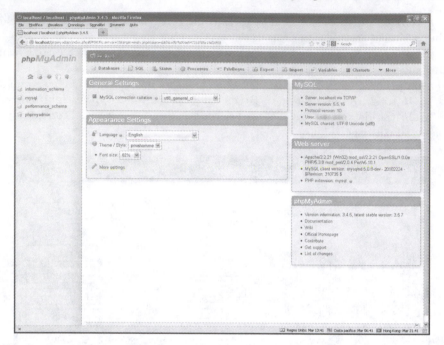

FIGURE 11.2
Browser window displaying the `localhost/phpmyadmin/` page.

11.3   Connecting R to MySQL

The `RMySQL` package provides R users with functions to connect to a MySQL database. Unfortunately the installation of such a package is usually not as straightforward as the typical installation of an R package (as shown in Chapter 2), especially under Windows. For this reason we provide an alternate way to connect to MySQL, by using the `RODBC` package.

Readers who plan to make extensive use of MySQL connections are encouraged to make the necessary efforts for installing `RMySQL`,[4] since that will have a much better performance. However, in the sections that follow, we provide alternate code for those working with `RODBC`.

11.3.1   Connecting using package `RMySQL`

Once `RMySQL` is installed, the R function `dbConnect` creates the connection. The `user` and `pwd` arguments indicate the user name and the password for

---

[4]Some details on how to get `RMySQL` to work under Windows can be found at `tinyurl.com/RMySQL1` and `tinyurl.com/RMySQL2`.

accessing the MySQL database (if they have been specified when the database was created), while `dbname` indicates the default database to which R will be connected (in our case `RBaseball`, created in Section 11.2). Note that the connection is assigned to an R object (`conn` here), as it will be required as an argument by several functions.

```
library(RMySQL)
conn <- dbConnect(MySQL(), user='username', pwd='password'
  , dbname='RBaseball')
```

To remove a connection one uses `dbDisconnect(conn=conn)`, where the argument `conn` requires an open connection.

11.3.2   Connecting using Package `RODBC`

The package `RODBC` provides an alternate way for connecting to a MySQL database. By making use of Open Database Connectivity (ODBC), `RODBC` provides a single solution to access databases from several DBMS.[5]

The first step is to set up an ODBC connection. ODBC is DBMS-independent because it uses drivers as "translation layers" between the application and the DBMS. To enable ODBC connection to MySQL (or any other DBMS) an appropriate driver is required. On the MySQL website the necessary driver can be downloaded at `dev.mysql.com/downloads/connector/odbc`.

Once the appropriate driver has been downloaded and installed, the subsequent required step is the configuration of a connector. The MySQL website provides detailed instructions on how to perform the task according to one's operating system: the relevant URL is `dev.mysql.com/doc/refman/5.0/en/connector-odbc-configuration.html`.

Here we illustrate the process for the configuration on a Windows XP machine, creating a connector to the `RBaseball` database created in Section 11.2.

1. From the Windows *Start* menu click on the *Control Panel*.

2. If the Control Panel is displayed in `Classic View`, click on *Administrative Tools*. If the Control Panel is displayed in *Category View* click on *Performance and Maintenance* and then on *Administrative Tools*.

3. Click on *Data Sources (ODBC)*.

4. In the *ODBC Data Source Administrator* box, click on *Add...* to add a new data source.

5. In the *Create New Data Source* box select the MySQL ODBC driver.

---

[5]A number of packages similar to `RMySQL` exist for connections to other DBMS systems: `RPostgreSQL`, `ROracle`, and `RSQLite` are a few of them.

6. Set the parameters in the *MySQL Connector/ODBC Data Source Config-uration* box as follows:

   (a) Enter a *Data Source Name* (DSN): in our case we choose `RBaseball`.

   (b) Optionally add a *Description*.

   (c) Enter `localhost` in the *Server* text box and leave the default `Port` value.

   (d) Enter a *User* and a *Password* for the connection.

   (e) From the *Database* popup menu, select the `RBaseball` database.

7. Click OK and the DSN is saved.[6]

Once the ODBC connector has been configured, the database can be accessed from R by use of the function `odbcConnect`. The argument `dsn` is the name of the data source used when creating the ODBC connection (point 6a in the previous subsection), while `uid` and `pwd` need to be provided if a user name and a password have been specified (6d) when setting the ODBC connection.

```
library(RODBC)
conn <- odbcConnect(dsn="RBaseball", uid="user"
  , pwd="password")
```

To close the connection in R, one types

```
odbcClose(channel=conn)
```

11.4   Filling a MySQL Game Log Database from R

The game log data files are currently available at the Retrosheet Web page `www.retrosheet.org/gamelogs/index.html`. By clicking on a single year one obtains a compressed (zip) file containing a single text file of the season's game logs. Here we first create a function for loading a season of game logs into R, then show how to append the data into a MySQL table. Finally code is provided which loops through several seasons, downloading the game logs from Retrosheet and appending them to the MySQL table.

---

[6]The *Test* button can be used to make sure the connection has been successfully configured.

11.4.1 From Retrosheet to R

The `load.gamelog` function, shown below, inputs the season number and performs the following operations:

1. Downloads the zip file of the season from the Retrosheet game log Web page.

2. Extracts the text file contained in the downloaded zip file.

3. Reads the text file into R.

4. Removes both the compressed and the extracted files.

```
load.gamelog <- function(season){
  download.file(
    url=paste("http://www.retrosheet.org/gamelogs/gl", season
      , ".zip", sep="")
    , destfile=paste("gl", season, ".zip", sep="")
  )
  unzip(paste("gl", season, ".zip", sep=""))
  gamelog <- read.table(paste("gl", season, ".txt", sep="")
    , sep=",", stringsAsFactors=F)
  file.remove(paste("gl", season, ".zip", sep=""))
  file.remove(paste("gl", season, ".txt", sep=""))
  gamelog
}
```

After the function has been read into R, one season of game logs (for example, the year 2012) is inputted into R by typing the command:

```
gl2012 <- load.gamelog(2012)
```

Since game log files downloaded from Retrosheet do not have column headers, the resulting `gl2012` data frame has column names assigned by default by R. They can be replaced with meaningful names stored in the `game_log_header.csv` file as done elsewhere in this book.

```
glheaders <- read.csv("retrosheet/game_log_header.csv")
names(gl2012) <- names(glheaders)
```

11.4.2 From R to MySQL

Using a ODBC connection, the `gl2012` data frame currently loaded in R can be transferred to the `RBaseball` MySQL database. In the lines that follow, the `RODBC` package is loaded, a connection is set to the `RBaseball` database, and finally the `sqlSave` function is used to append the data to a table in the MySQL database.

```
library(RODBC)
conn <- odbcConnect("RBaseball")
sqlSave(channel=conn, dat=gl2012, tablename="gamelogs"
  , append=TRUE, rownames=FALSE)
```

Here are some notes on the arguments in the `sqlSave` function:

- The `channel` argument requires an open connection; here the one that was previously set (`conn`) is specified.

- The `dat` argument requires the name of the R data frame to be appended to the table in the MySQL database.

- The `tablename` argument requires a string indicating the name of the table (in the database) where the data are to be appended.

- Setting `append` to `TRUE` indicates that, should a table by the name `"gamelogs"` already exist, data from `gl2012` will be appended to the table. If `append` is set to `FALSE` the table `"gamelogs"` (if it exists) will be overwritten.

- The `rownames=FALSE` argument indicates the row names will not be appended.

The R code using `RMySQL` is similar to the `RODBC` code, but requires significantly less time for appending data. The following code is preferable if one has successfully installed the `RMySQL` package.

```
library(RMySQL)
conn <- dbConnect(MySQL(), user='username'
  , dbname='RBaseball')
dbWriteTable(conn, name="gamelogs", value=gl2012
  , append=TRUE, row.names=F)
```

In the previous section, code was provided for appending one season of game logs into a MySQL table. However, we have demonstrated in previous chapters that it is straightforward to use R to work with a single season of game logs. To fully appreciate the advantages of storing data in a DBMS, a MySQL table will be populated with game logs going back through baseball history. With a historical database and an R connection, we demonstrate the use of R to perform analysis over multiple seasons.

A new function `appendGameLogs` is written which loops through a specified set of years (potentially from 1871 to the present), downloads the files from Retrosheet,[7] and appends the data to the `gamelogs` table in the `RBaseball`

---

[7]The downloading of data from Retrosheet is performed by the previously presented `load.gamelog` function, thus the reader has to make sure said function is loaded for the code in this section to work.

MySQL database.[8] The whole process may take several minutes. If one is not interested in downloading files dating back to 1871, seasons from 1995 are sufficient for reproducing the example of the next section.

The function `appendGameLogs` takes the following parameters as inputs:

- `start` and `end` indicate the seasons one wants to download from Retrosheet and append to the MySQL database. By default the function will work on seasons from 1871 to 2012.

- `connPackage` provides the user with the option of selecting whether to use the `RODBC` or the `RMySQL` package for performing the work. Note that `RMySQL` works faster and is the preferred choice.

- `headersFile` points to the full path where the file containing the game log headers is stored.

- `dbTableName` specifies the name of the table in the MySQL database where the data are to be uploaded.

- ther parameter inputs as required by either the `odbcConnect` or the `dbConnect` functions, such as the DSN name, the user name, and the password to access the database.[9]

```
appendGameLogs <- function(
  start=1871, end=2012, connPackage="RODBC"
  , headersFile="retrosheet/game_log_header.csv"
  , dbTableName="gamelogs", ...
){
  require(package=connPackage, character.only=TRUE)
  glheaders <- read.csv(headersFile)
  if(connPackage == "RMySQL"){
    conn <- dbConnect(MySQL(), ...)
  } else {
    conn <- odbcConnect(...)
  }
  for(season in start:end){
    print(paste(Sys.time(), "working on season:", season))
    flush.console()
    gamelogs <- load.gamelog(season)
    glheaders <- read.csv(headersFile)
    names(gamelogs) <- names(glheaders)
    gamelogs$GAME_ID <- paste(gamelogs$HomeTeam, gamelogs$Date
```

---

[8]The proposed code actually allows the reader to specify both a database and a table of choice.

[9]The three-dots (...) construct is used here for allowing the user to specify additional arguments to the `appendGameLogs` function as needed by the functions called inside it (either `dbConnect` or `odbcConnect`).

```
                , gamelogs$DoubleHeader, sep="")
    gamelogs$YEAR_ID <- substr(gamelogs$Date, 1, 4)
    if(connPackage == "RMySQL"){
      dbWriteTable(conn, name=dbTableName, value=gamelogs
                   , append=TRUE, row.names=F)
    } else {
      sqlSave(conn, dat=gamelogs, tablename=dbTableName
        , append=TRUE, rownames=FALSE)
    }
  }
}
```

The following lines of code demonstrate the use of the function appendGameLogs both with RODBC and RMySQL.

```
appendGameLogs(start=1871, end=1872, connPackage="RODBC"
  , dsn="rbaseball", uid="user", pwd="password")

appendGameLogs(start=1873, end=1874, connPackage="RMySQL"
  , dbname="rbaseball", user="user", password="password")
```

11.5 Querying Data from R

11.5.1 Introduction

Both RMySQL and RODBC provide functions for querying data stored in a MySQL database. The sqlFetch function from RODBC, for example, allows the import of a full MySQL table into an R data frame. However, the purpose of having data stored in a MySQL database is to selectively import data into R for particular analysis; one selectively imports data by querying one or more tables of the database.

For example, suppose one is interested in comparing the attendance of the two Chicago teams by day of week since the 2000 season. The following code retrieves the raw data in R.

```
library(RODBC)
conn <- odbcConnect("RBaseball")
chiAttendance <- sqlQuery(conn, "
                    select date, hometeam, dayofweek
                      , attendence
                    from gamelogs
                    where date > 20000101
                    and hometeam in ('CHN', 'CHA')
                    ")
```

```
head(chiAttendance)
  hometeam dayofweek attendance
1    CHN      Thu       55000
2    CHN      Mon       38655
3    CHN      Wed       26838
4    CHN      Thu       20152
5    CHN      Fri       21324
6    CHA      Fri       38912
```

The `sqlQuery` function provides the database query. Its arguments in this function are the connection handle established in the previous line (`conn`) and a string consisting of a valid SQL statement. Readers familiar with SQL will have no problem in understanding the meaning of the query. For those unfamiliar with SQL, we present here a brief explanation of the purpose of the query, inviting anyone who is interested in learning about the language to look for the numerous resources devoted to the subject.

The first row in the SQL statement indicates the columns of the table that are to be *select*-ed (in this case `date`, `hometeam`, `dayofweek`, and `attendance`). The second line states *from* which table they have to be retrieved (`gamelogs`). Finally, the *where* clause specifies conditions for the rows that are to be retrieved: the `date` has to be greater than 20000101 *and* the value of `hometeam` has to be one of `CHN` and `CHA`.

The same task can be performed with the following code if one wants to use RMySQL rather than RODBC.

```
library(RMySQL)
conn <- dbConnect(MySQL(), user='username', pwd='password'
  , dbname='RBaseball')
chiAttendance <- dbGetQuery(conn, "
                    select date, hometeam, dayofweek
                      , attendence
                    from gamelogs
                    where date > 20000101
                    and hometeam in ('CHN', 'CHA')
                    ").
```

The comparison between attendance at the two Chicago ballparks is displayed in Figure 11.3. The plot has been obtained by first setting as NA the games reporting attendance values of zero,[10] then transforming the `dayofweek` column into a factor ordered from `Sun` to `Sat` and calculating the average attendance by day of week. Finally the `xyplot` function from the `lattice` package has been used to actually draw the plot.

---

[10]In case of single admission doubleheaders (i.e., when two games are played on the same day and a single ticket is required for attending both) the attendance is reported only for the second game, while it is set at zero for the first.

```
chiAttendance$attendence <- ifelse(chiAttendance$attendence == 0
    , NA, chiAttendance$attendence)
chiAttendance$dayofweek <- factor(chiAttendance$dayofweek
    , levels=c("Sun", "Mon", "Tue", "Wed", "Thu", "Fri", "Sat"))
avgAtt <- aggregate(attendence ~ hometeam + dayofweek
    , data=chiAttendance, FUN=mean)

library(lattice)
xyplot(attendence ~  dayofweek, data=avgAtt
        , groups=hometeam
        , pch=c("S", "C"), cex=2, col= "black"
        , xlab="day of week"
        , ylab="attendance"
        )
```

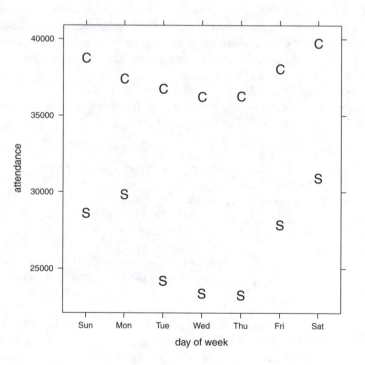

FIGURE 11.3
Comparison of attendance by day of the week on games played at home by
the Cubs (C) and the White Sox (S).

11.5.2   Coors Field and run scoring

As an example of accessing multiple years of data, we explore the effect of Coors Field (home of the Colorado Rockies in Denver) on run scoring through the years. Coors Field is a peculiar ballpark because it is located at an altitude of about one mile over the sea level. Coors Field has thinner air than other stadiums, allowing both batted balls to travel farther and making the curveballs delivered by pitchers "flatter."

As a first step, we use the function `dbConnect` to connect R to the MySQL database containing the game logs. In the use of this function, remember to adjust the argument values wherever necessary.

```
library(RMySQL)
conn <- dbConnect(MySQL(), user='username', pwd='password'
  , dbname='retrosheet', host='localhost')
```

Then, by using SQL language in the function `dbGetQuery`,[11] data is retrieved for the games played by the Rockies, either at home or on the road, since 1995, the year they moved to Coors Field.

```
rockies.games <- dbGetQuery(conn, "select year_id, date, parkid
                          , visitingteam, hometeam
                          , visitorrunsscored as awR
                          , homerunsscore as hmR
                         from gamelogs
                        where (hometeam='COL'
                       or visitingteam='COL')
                      and year_id > 1994")
```

The game data is conveniently stored in the `rockies.games` data frame and can be explored with R commands. The sum of runs scored in each game is computed by adding the runs scored by the home team and the visiting team. A new column `coors` is added indicating whether the game was played at Coors Field.[12]

```
rockies.games$runs <- rockies.games$awR + rockies.games$hmR
rockies.games$coors <- (rockies.games$parkid == "DEN02")
```

The offensive output by the Rockies and their opponents is compared at Coors and other ballparks by a graph constructed using the `ggplot2` package introduced in Chapter 6. The resulting visualization is displayed in Figure 11.4.

```
library(ggplot2)
ggplot(aes(x=year_id, y=runs, linetype=coors), data=rockies.games) +
```

---

[11] The keyword `as` in SQL has the purpose of assigning different names to columns. Thus `visitorrunsscored as awR` tells SQL that, in the results returned by the query, the column `visitorrunsscored` will be named `awR`.

[12] Retrosheet code for Coors Field is `DEN02`. A list of all ballparks codes is available at `www.retrosheet.org/parkcode.txt`.

```
stat_summary(fun.data="mean_cl_boot") +
xlab("season") +
ylab("runs per game (both teams combined)") +
scale_linetype_discrete(name="location"
  , labels=c("other parks", "Coors Field"))
```

The `stats_summary` layer is used in `ggplot2` to summarize the y values at every unique value of x. The `fun.data` argument lets the user specify a summarizing function; in this case `"mean_cl_boot"` implements a nonparametric bootstrap procedure for obtaining confidence bands for the population mean. The output resulting from this layer are the vertical bars appearing for each data point. The `scale_linetype_discrete` layer is used for labeling the series (`name` argument) and assigning a name to the legend (`labels`).

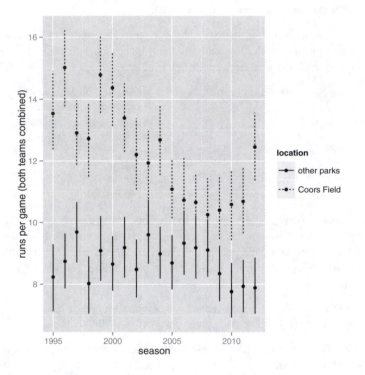

FIGURE 11.4
Comparison of runs scored by the Rockies and their opponents at Coors Field and in other ballparks.

From Figure 11.4 one notices how Coors Field has been an offense-friendly park, boosting run scoring by as high as six runs per game. However the effect of the Colorado ballpark has somewhat decreased in the new millennium, displaying minimal differences in the 2006-2008 period. One reason for Coors becoming less of an extreme park is the installation of a humidor. Since the

2002 season, baseballs have been stored, prior to each game, in a room at a higher humidity, with the intent of compensating for the unusual natural atmospheric conditions.[13]

11.6 Baseball Data as MySQL Dumps

Section 11.4.1 illustrated populating a MySQL database from within R using an example of creating a table of Retrosheet game logs. Several so-called *SQL dumps* are available online for creating and filling databases with baseball data. SQL dumps are simple text files (featuring a `.sql` extension) containing SQL instructions for creating and filling SQL tables.

11.6.1 Lahman's database

Sean Lahman provides his historical database of seasonal stats in several formats. So far in this book we have used the comma-delimited version, consisting of several `.csv` files. However, the database is also available as a SQL dump, which can be downloaded from `seanlahman.com/baseball-archive/statistics/` (look for the *SQL version*). The downloaded file is a zip archive, which needs to be decompressed. The resulting extracted files include one file dedicated to instructions (`readme 2012.txt` in the version downloaded at the time of this writing) and the dump file (`lahman2012.sql.`)

There are several options for having the dump file perform its task, including typing commands in the OS shell or importing the file via the phpAdmin panel. It is beyond the scope of this book to illustrate said processes, thus here we provide a command to obtain the desired result from within R by use of the `shell` function which invokes the Command Prompt (in Windows). In order for the following code to work, the MySQL service should be running[14] (see Section 11.2 and Figure 11.1).

```
shell("mysql < C:\\Baseball\\Datadumps\\lahman2012.sql -u username
  -p password")
```

Note that the SQL dump file creates a new database named `lahman`.

---

[13]For a detailed analysis of the humidor effects, see this article by Alan Nathan, Professor Emeritus of Physics at the University of Illinois at Urbana-Champaign: `www.baseballprospectus.com/article.php?articleid=13057`.

[14]Also, make sure to change the directory containing the `.sql` file and the user name and the password as appropriate.

11.6.2   Retrosheet database

In Appendix A we provide R code to download Retrosheet files and transform them in formats easily readable by R. By slightly adapting the code provided in Section 11.4.2, one can also append them to a MySQL database. Jeff Zimmerman provides SQL dumps of the Retrosheet database at the Baseball Heat Maps website. At the URL www.baseballheatmaps.com/retrosheet-database-download/ one has several download options, including the full Retrosheet database and downloads by decade.

Once one has downloaded the desired file and extracted the content from the zip archive, a command similar to the one previously shown for the Lahman's database will create the tables in MySQL. Note that the SQL dump provided by Zimmerman does not contain code to create (and use) a new database, thus one has to specify the name of the database where operations are to be performed in the following code[15] we have indicated the RBaseball database after the password parameter.

```
shell("mysql < C:\\Baseball\\Datadumps\\2010s_Retrosheet.sql
 -u username -p password RBaseball")
```

Zimmerman's SQL dump creates a number of lookup tables, useful for decoding information such as the type of batted ball, the base situation, and so on. Details on coded values are provided in Appendix A.

11.6.3   PITCHf/x database

Zimmerman also provides downloads, updated daily during the season, of PITCHf/x data at www.baseballheatmaps.com/pitch-fx-download/. Similarly to what was shown in the previous section, after downloading the zip archive and extracting its content to a folder of choice, the following command in R will provide the reader with an up-to-date MySQL PITCHf/x database.

```
shell("mysql < C:\\Baseball\\Datadumps\\pbp2.sql
 -u username -p password RBaseball")
```

Zimmerman's data dump comes with a handful of tables. The bulk of PITCHf/x data is contained in the pitches table, which features most of the information the reader was presented with in Chapter 6, when the verlander and cabrera data frames were introduced. A more detailed explanation on the columns featured in the pitches table can be found in Appendix B, particularly in Section B.4.2.

Similarly, some details on the contents of the atbats table are provided in Sections B.4.1 and B.4.3. Whenever ones needs to combine information contained in pitches with information contained in atbats, the ab_id is the common column to be used for merging purposes.

---

[15]For the code sample we have downloaded Retrosheet data for the 2010-present period.

The SQL dump also contains a `games` table, featuring general information on the games, which can be linked to `atbats` with the common `game_id` column. Finally a few reference tables provide details on players, umpires, teams, and pitch types.

11.7   Calculating Basic Park Factors

Park factors (usually abbreviated as PF) have been used for decades by baseball analysts as a tool for mitigating the effect of the ballpark when assessing the value of players. Park factors have been calculated in several ways and in this section we illustrate a very basic approach, focusing on year 1996, one of the most extreme seasons for Coors Field, as displayed in Figure 11.4.

In the explanations that will follow, we presume the reader has Retrosheet data for the 1990s in his database.[16]

11.7.1   Loading the data in R

As a first step, we connect to the MySQL database (in this example via an ODBC connection) and retrieve the desired data: by using SQL language we *select* the columns containing the home and away teams and the event code *from* the `events` table, keeping only the rows *where* the year is 1996 *and* the event code corresponds to one indicating a batted ball (see Appendix A). The results of the query are stored in the `hrPF` R data frame.

```
# Connect to MySQL database
options(stringsAsFactors=F)
library(RODBC)
conn <- odbcConnect("RBaseball")

# get data
hrPF <- sqlQuery(conn, "
            select away_team_id, home_team_id, event_cd
            from events
            where year_id=1996
            and event_cd in (2, 18, 19, 20, 21, 22, 23)
            ")
```

----

[16]Refer to Section 11.6.2 for performing the necessary steps to get the data into a MySQL database.

11.7.2   Home run park factor

A ballpark can have different effects on the various players performance statistics. The unique configuration of Fenway Park in Boston, for example, enhances the likelihood of a batted ball to become a double, especially flyballs to left, which often carom off the Green Monster. On the other hand, home runs are rare on the right side of Fenway Park due to the unusually long distance of the right-field fence from home plate.

In this example we explore the stadium effect on home runs in 1996, calculating *park factors* for home runs. To begin, a new column `event_fl` is created which indicates, for every row in the `hrPF` data frame, the occurrence of a home run.

```
hrPF$event_fl <- ifelse(hrPF$event_cd == 23, 1, 0)
```

Next the frequency of home runs per batted ball is computed for all MLB teams both at home and on the road. The two resulting data frames are merged and the first column of the newly obtained data frame `evCompare` is renamed to `team_id`.

```
evAway <- aggregate(event_fl ~ away_team_id, data=hrPF
    , FUN=mean)
evHome <- aggregate(event_fl ~ home_team_id, data=hrPF
    , FUN=mean)
evCompare <- merge(evAway, evHome, by.x="away_team_id"
    , by.y="home_team_id", suffixes=c("_A", "_H"))
names(evCompare)[1] <- "team_id"

head(evCompare)
  team_id event_fl_A event_fl_H
1     ANA 0.03808699 0.03053270
2     ARI 0.03013345 0.03935524
3     ATL 0.04524887 0.04268827
4     BAL 0.03729904 0.04293456
5     BOS 0.04663212 0.03374167
6     CHA 0.04045377 0.05190166
```

Park factors are typically calculated so that the value 100 indicates a neutral ballpark (one which has no effect on the particular statistic) while values over 100 indicate playing fields that increase the likelihood of the event (home run in this case) and values under 100 indicate ballparks that decrease the likelihood of the event.

The 1996 home run park factors are obtained with the first line of the following code. The resulting data frame `evCompare` is ordered in descending order by `PF`. The `head` function is used to display the most HR-friendly parks and the `tail` function displays the least-friendly home run parks. Coors Field is at the top of the HR-friendly list, displaying an extreme value of 158 – this park boosted home run frequency by over 50% in 1996. At the other end of the spectrum, in 1996 was Dodgers Stadium in Los Angeles, featuring a home run park factor of 71.

```
evCompare$PF <- round(100 * evCompare$event_fl_H
  / evCompare$event_fl_A)
evCompare <- evCompare[order(-evCompare$PF),]
head(evCompare)
   team_id event_fl_A event_fl_H  PF
9      COL 0.03405430 0.05383393 158
4      CAL 0.03870425 0.04831843 125
1      ATL 0.03228680 0.03717311 115
3      BOS 0.03849444 0.04425145 115
10     DET 0.04571550 0.05058280 111
6      CHN 0.03741794 0.04066623 109
tail(evCompare)
   team_id event_fl_A event_fl_H PF
13     KCA 0.03380102 0.02905167 86
8      CLE 0.04402386 0.03722397 85
5      CHA 0.04240652 0.03486662 82
12     HOU 0.03441600 0.02723649 79
19     NYN 0.03634809 0.02888087 79
14     LAN 0.03595703 0.02561144 71
```

11.7.3 Assumptions of the proposed approach

The proposed approach to calculating park factors makes several simplifying assumption. The first assumption is that the home team and the ballpark are considered interchangeable. While that is true for most games, sometimes alternate ballparks have been used for particular games. For example, during the present season, the Oakland Athletics played their first six home games in Cashman Field (Las Vegas, NV) while renovations at the Oakland-Alameda County Coliseum were being completed, and the San Diego Padres had a series of three games against the New York Mets at Estadio de Beisbol Monterrey in Mexico as a marketing move by MLB.[17]

Another assumption of the proposed approach is that a single park factor is appropriate for all players, without considering how ballparks might affect some categories of players differently. Asymmetric outfield configurations, in fact, cause playing fields to have unequal effects on right-handed and left-handed players. For example, the aforementioned Green Monster in Boston, being situated in left field, comes into play more frequently when right-handed batters are at the plate; and the recent new version of the Yankee Stadium

---

[17]A list of games played in alternate sites is displayed on the Retrosheet website at the url `www.retrosheet.org/neutral.htm`.

has seen left-handed batters take advantage of the short distance of the right field fence.

Finally, the proposed park factors (as well as most published versions of park factors) essentially ignore the players involved in each event (in this case the batter and the pitcher). As teams rely more on the analysis of play-by-play data, they typically adapt their strategies to accommodate the peculiarities of ballparks. For example, while the diminished effect of Coors Field on run scoring displayed in Figure 11.4 is mostly attributable to the humidor, part of the effect is certainly due to teams employing different strategies when playing in this park. For example, teams can use pitchers who induce a high number of groundballs that are less impacted by the effect of the rarified air.

11.7.4   Applying park factors

In the 1996 season, four Rockies players hit 30 or more home runs – Andres Galarraga led the team with 47, followed by Vinnie Castilla and Ellis Burks (tied at 40) and Dante Bichette (31). Behind them Larry Walker had just 18 home runs, but with a very limited playing time due to injuries. In fact, Walker's HR/AB ratio was second to only Galarraga's. Their offensive output was certainly boosted by playing 81 of their games in Coors Field. Using the previously calculated park factors, one can estimate the number of home runs Galarraga would have hit in a neutral park environment.

We begin with retrieving from the MySQL database every Galarraga's 1996 plate appearance ending with a batted ball, and a column `event_fl` is defined indicating a value of 1 for home runs and 0 otherwise, as was done in Section 11.7.2.

```
# Connect to MySQL database
options(stringsAsFactors=F)
library(RODBC)
conn <- odbcConnect("rbaseball")

# Get Galarraga's plate appearances
andres <- sqlQuery(conn, "
                select away_team_id, home_team_id, event_cd
                from events
                where year_id=1996
                and event_cd in (2, 18, 19, 20, 21, 22, 23)
                and bat_id='galaa001'
                ")

# identify HRs
andres$event_fl <- ifelse(andres$event_cd == 23, 1, 0)
```

Then the previously calculated park factors are added to the `andres` data frame. This is done by merging data frames `andres` and `evCompare`, using the columns `home_team_id` and `team_id` as the merging columns. Using the merged data frame `adresPF`, we calculate the mean park factor for Galarraga's plate appearances.

```
andres <- merge(andres, evCompare[,c("team_id", "PF")]
                , by.x="home_team_id"
                , by.y="team_id")
andresPF <- mean(andres$PF)
andresPF
[1] 129.1384
```

The compounded park factor for Galarraga, derived from the 252 batted balls he had at home and the 225 he had on the road (ranging from 9 in Dodgers Stadium in Los Angeles to 23 at the Astrodome in Houston), indicates Andres had his home run frequency increased by an estimated 29% relative to a neutral environment. In order to get the estimate of home runs in a neutral environment, we divide Galarraga's home runs by his average home run park factor multiplied by 100.

```
47 / andresPF * 100
[1] 36.39507
```

According to our estimates, Galarraga's benefit from the ballparks he played in (particularly his home Coors Field) amounted to roughly 47 - 36 = 11 home runs in the 1996 season.

11.8 Further Reading

Chapter 2 of Adler (2006) has detailed instructions on how to obtain and install MySQL, and on how to set up an historical baseball database with Retrosheet data. Hack #56 (Chapter 5 of the same book) provides SQL code for computing and applying Park Factors.

MySQL reference manuals are available in several formats at `dev.mysql.com/doc/` on MySQL website. The HTML online version features a search box which allows users to quickly retrieve pages pertaining to specific functions.

11.9 Exercises

1. (**Runs Scored at the Astrodome**)

   (a) Using either the `sqlQuery` function from the `RODBC` package or the dbGetQuery function from the RMySQL package, select games featuring the Astros (as either the home or visiting team) during the years when the Astrodome was their home park (i.e., from 1965 to 1999).

(b) Draw a plot to visually compare through the years the runs scored (both teams combined) in games played at the Astrodome and in other ballparks.

2. (**Astrodome Home Run Park Factor**)

(a) Select data from one season between 1965 and 1999. Keep the columns indicating the visiting team identifier, the home team identifier and the event code, and the rows identifying ball-in-play events. Create a new column which identifies whether an home run has occurred.

(b) Prepare a data frame containing the team identifier in the first column, the frequency of home runs per batted ball when the teams plays on the road in the second column, and the same frequency when the team plays at home in the third column.

(c) Compute home run park factors for all MLB teams and check how the domed stadium in Houston affected home run hitting.

3. (**Applying Park Factors to "Adjust" Numbers**)

(a) Using the same season selected for the previous exercise, obtain data from plate appearances (ending with the ball being hit into play) featuring one Astros player of choice. The exercise can either be performed on plate appearances featuring an Astros pitcher on the mound or an Astros batter at the plate. For example, if the selected season is 1988, one might be interested in discovering how the Astrodome affected the number of home runs surrendered by veteran pitcher Nolan Ryan (Retrosheet id: ryann001) or the number of home runs hit by rookie catcher Craig Biggio (id: biggc001).

(b) As shown in Section 11.7.4 merge the selected player's data with the Park Factors previously calculated and compute the player's individual Park Factor (which is affected by the different playing time the player had in the various ballparks) and use it to estimate a "fair" number of home runs hit (or surrendered if a pitcher was chosen).

4. (**Park Factors for Other Events**)

(a) Park Factors can be estimated for events other than Home Runs. SeamHeads.com Ballpark Database[18] for example features Park Factors for seven different events, plus it offers split factors according to batters' handedness. See for example the page for the Astrodome: www.seamheads.com/ballparks/ballpark.php?parkID=HOU02&tab=pf1.

---

[18]www.seamheads.com/ballparks/index.php

(b) Choose an event (even different from the seven shown at SeamHeads) and calculate how ballparks affect its frequency. As a suggestion, the reader may want to look at seasons in the '80s, when artificial turf was installed in close to 40% of MLB fields, and verify whether parks with concrete/synthetic grass surfaces featured a higher frequency of batted balls (home runs excluded) converted into outs.[19]

---

[19]SeamHeads provides information on the surface of play in each stadium's page. For example, in the previously mentioned page relative to the Astrodome, if one hovers the mouse over the ballpark name in a given season, a pop-up will appear providing information both on the ballpark cover and its playing field surface. SeamHeads is currently providing its ballpark database as a zip archive containing comma-separated-value (.csv) files that can be easily read by R: the link for downloading it is found at the bottom of each page in the ballpark database section.

12

Exploring Fielding Metrics with Contributed R Packages

CONTENTS

12.1   Introduction 283
12.2   A Motivating Example: Comparing Fielding Metrics 284
     12.2.1   Introduction 284
     12.2.2   The fielding metrics 285
     12.2.3   Reading an Excel spreadsheet (`XLConnect`) 286
     12.2.4   Summarizing multiple columns (`doBy`) 287
     12.2.5   Finding the most similar string (`stringdist`) 288
     12.2.6   Applying a function on multiple columns (`plyr`) 291
     12.2.7   Weighted correlations (`weights`) 291
     12.2.8   Displaying correlation matrices (`ellipse`) 292
     12.2.9   Evaluating the fielding metrics (`psych`) 293
12.3   Comparing Two Shortstops 294
     12.3.1   Reshaping the data (`reshape2`) 296
     12.3.2   Plotting the data (`ggplot2` and `directlabels`) 296
12.4   Further Reading 297
12.5   Exercises 298

12.1   Introduction

R packages have been introduced in Chapter 2 of this book. Here the evaluation of fielding is used as a motivating example to further illustrate the capabilities of a number of R packages. To use any of the packages described in this chapter, the package needs to be installed in R before loading the package with the `library` function. One installs a package by use of the `install.packages` function or the *Install Packages* button in RStudio's *Packages* tab.

12.2   A Motivating Example: Comparing Fielding Metrics

12.2.1   Introduction

The evaluation of defensive contribution is one of sabermetrics' holy grails. Fielding measures have vastly improved since the days when fielding percentage[1] was the only available metric for defensive evaluation. In evaluating fielding fifty years ago, one had to choose between using a very imperfect tool or ignoring fielding altogether. Big strides in evaluating defense have been possible due to the increasing availability of data. Retrosheet has made decades of play-by-play information accessible to the public, and companies like STATS, Baseball Info Solutions and Sportvision have been tracking, at increasing levels of detail and precision, the ball trajectories and the positioning of the fielders.

The quantification of defensive contribution is a subject which could fill the pages of an entire book. Recently, tax lawyer Michael A. Humphreys published a book *Wizardry - Baseball's All-Time Greatest Fielders Revealed* in which he provides an overview of the publicly known defensive metrics and proposes his own measure which allows the assessment of defensive value throughout the history of baseball.

In this chapter we introduce a handful of defensive metrics with the goal of using these measures to evaluate the fielding of shortstops during the 2009 season. Since the ratings for one of the metrics are available as a Microsoft Excel spreadsheet, we introduce the `XLConnect` package that provides functions for working with Excel files within R. A few shortstops played for different teams during the 2009 season, but we are interested in their defensive ratings for the full season. The package `doBy` allows for summarizations of data frames over multiple columns and this package is helpful for describing defensive ratings split by teams.

It can be problematic combining data from two different sources with no common identifier and different spellings of the players' names. In this work, it is helpful to use the `stringdist` package and its functions devised for the comparison of strings. Then the function `colwise` function from the `plyr` package is shown helpful in applying the same function over multiple columns of a data frame.

Several useful functions for working with correlations will be demonstrated. After correlations are computed using the `cor` function from the `stats` package (which comes installed with the basic version of R), a special function from the `weights` package computes correlations that allow for the assignment of

---

[1] Fielding percentage is defined as the ratio between a fielder's successful chances (putouts + assists) and his total chances (putouts + assists + errors).

different weights to the observations in the data frame. Correlations can be attractively displayed by use of functions in the `ellipse` package.

12.2.2 The fielding metrics

The majority of defensive metrics are derived through basically the same process. Each measure compares the number of successful plays of a fielder with the expected number an average fielder would make with the same opportunities and translates the extra (fewer) plays made into runs saved (cost) to the team. Given this common framework, defensive metrics differ on the data used, the computation of the expected number of plays, and the method in which plays are converted to runs. Six fielding metrics are compared in this chapter; we provide a brief introduction on the history of each measure and how it is defined.

- **Ultimate Zone Rating (UZR)**, developed by Mitchel Lichtman, is based on proprietary data collected by Baseball Info Solutions (BIS). For each batted ball, the system estimates the probability of getting an out given several parameters such as the speed and trajectory of the batted ball, the handedness of the batter, the base/out situation, and the ballpark.

- **Defensive Runs Saved (DRS)** was introduced in *The Fielding Bible* by John Dewan, the owner of Baseball Info Solutions. This system estimates the number of extra (fewer) plays made by a fielder given a set of parameters defining the batted ball, including the direction, the batted ball type (groundball, flyball, line drive, pop up), and the type of contact (soft, medium, hard). DRS uses proprietary BIS data, obtained with the work of so-called videoscouts–i.e., people entering information in databases while watching recorded broadcasts of games.

- **Total Zone (TZL)**, developed by Sean Smith, makes use of the freely available Retrosheet data. It determines the fielder responsible for a batted ball using (whenever available) the batted ball type and location. Since that data is not available in every season covered by Retrosheet, Smith uses information about batter handedness, pitcher's ground ball/fly ball tendency, and the fielder ultimately collecting the batted ball as proxies to designate the responsible fielder.

- **Revised Zone Rating (RZR)** divides the field of play into several zones and assigns the responsibility of fielding a batted ball to one fielder depending on the zone where the ball is hit. In its basic version, Zone Rating (ZR) is simply the ratio of balls in a fielder's zone converted into outs. In its revised form, a bonus is added to account for the plays made by fielders on balls outside of their zones.

- **Fan Scouting Report (FSR)** is a *subjective* defensive evaluation system. Its numbers are based on the results of a yearly survey among baseball fans conducted by analyst and blogger Tom Tango.

- ***Wizardry*'s Runs (Runs)** are taken from Humphreys' book and are derived from seasonal data available in the Lahman's database. One advantage of this system is that it is possible to calculate fielding ratings throughout baseball history.

As a rule of thumb for the counting metrics (all of the above, except RZR), one can identify the players who are credited with saving 15 or more runs as extremely good ones and, conversely, as awful defenders those who are estimated costing 15 or more runs to their teams. For RZR, where an average fielder usually posts a rate around .835, the best players are found at the .940 mark, while the worst ones record values close to .700.

12.2.3   Reading an Excel spreadsheet (XLConnect)

The appendices of Humphreys (2011) are available online on the book's website.[2] Appendix C of Humphreys (2011) consists of nine Excel spreadsheets containing historical defensive contributions calculated by the author's method. In the examples that follow, we will make use of the shortstops' spreadsheet measures, thus readers who are following this code on R should download this spreadsheet and place it in a folder of choice.

A few contributed packages are helpful for reading Excel files into R including the **XLConnect** package. In the following R script, the package is loaded with the `library` function, and the `loadWorkbook` function is used to read in the Excel file of spreadsheet measures; the workbook in R is named `xlwzr`. The `readWorksheet` function is used to read the first sheet of `xlwzr` into a data frame with name `wzr`, starting from row number seven (with the `startRow` argument) to skip the header lines. The final lines of the code use the `subset` function to create a new data frame restricting `Year` to 2009, and a new variable `Name` is created which combines the first and last names of the players.

```
library(XLConnect)
xlwzr <- loadWorkbook("Appendix_C_Shortstop.xls")
wzr <- readWorksheet(xlwzr, sheet=1, startRow=7)
wzr <- subset(wzr, Year == 2009)
wzr$Name <- paste(wzr$First, wzr$Last)
head(wzr)
    Year L T.mcs   First   Last Pos       IP       Runs
236 2009 N   FLA Alfredo Amezaga  SS  42.0000  1.0780581
275 2009 A   BAL  Robert  Andino  SS 478.3333 -0.7244126
284 2009 A   TEX   Elvis  Andrus  SS 1238.0000  5.0929674
417 2009 A    KC    Mike  Aviles  SS 269.3333  5.0588753
```

---

[2]www.oup.com/us/companion.websites/9780195397765/appendices/?view=usa.

```
425 2009 A    LA   Erick   Aybar  SS 1189.3333 11.1041201
571 2009 N   STL   Brian  Barden  SS   27.0000  0.5520377
         v.Tm            Name
236  1.514699 Alfredo Amezaga
275 -3.394066  Robert Andino
284 -4.307379    Elvis Andrus
417  3.166190    Mike Aviles
425  7.698325    Erick Aybar
571  0.180907   Brian Barden
```

The data frame `wzr` contains, in order, the year, the league, and the team for which the shortstop played, the player's first and last name, his position, the number of innings played, his defensive runs, the difference between his defensive runs and those of the rest of his team at the same position, and the newly created variable with the full player's name.

12.2.4   Summarizing multiple columns (doBy)

Some of the players in the `wzr` data frame have multiple entries, such as Orlando Cabrera who played shortstop for multiple teams in 2009.

```
subset(wzr, Name == "Orlando Cabrera")
     Year L T.mcs  First    Last Pos       IP        Runs
1650 2009 A   OAK Orlando Cabrera  SS 887.6667 -12.7156058
1651 2009 A   MIN Orlando Cabrera  SS 501.0000  -0.7264558
         v.Tm            Name
1650 -4.554030 Orlando Cabrera
1651 -2.043441 Orlando Cabrera
```

To obtain the cumulative seasonal data for all the players, one needs to sum the `IP`, the `Runs`, and the `v.Tm` columns for players who played in multiple teams. The `aggregate` function allows one to perform this task for one column at a time; with three applications of `aggregate`, the result is three separate data frames that need to be merged using the `merge` function.

An alternative approach to `aggregate` uses the `doBy` package which features functions for groupwise calculations. One particularly useful function is `summaryBy` which works similarly to `aggregate`, but allows for the use of multiple columns on the left side of the aggregating formula. In the following R code, the `doBy` package is loaded and the `summaryBy` is used to sum[3] on the three columns as desired, grouping the data by the `Name` column. The `keep.names` arguments is set to `TRUE` in order to maintain the original names for the columns.

---

[3]The `FUN` argument in `summaryBy` also accepts a vector, thus multiple functions can be passed, and thus performed, at once. For example with `FUN=c(mean, sd)` one simultaneously computes the mean and the standard deviation on the variables specified on the left side of the formula. When more than one function is passed to `FUN`, it is not possible to set `keep.names` as `TRUE`.

```
library(doBy)
wzr <- summaryBy(IP + Runs + v.Tm ~ Name, data=wzr, FUN=sum,
  keep.names=TRUE)
```

To confirm that this function is operating correctly, the first few lines of the
wzr data frame are displayed.

```
head(wzr)
          Name       IP       Runs       v.Tm
1    A. Hernandez 289.6667 -3.2744424 -0.6064142
2     Aaron Miles  44.0000  0.0488766 -0.3746494
3   Adam Everett 942.6667  2.7007132 -2.3298411
4   Adam Rosales  33.0000  0.8441731  0.7099337
5 Alberto Callaspo   2.0000 -0.2630282 -0.2770827
6 Alberto Gonzalez 279.3333 -4.9981058 -4.6320992
```

12.2.5   Finding the most similar string (stringdist)

Data for several other defensive metrics can be downloaded from the Fan-
Graphs website.[4] Pointing the mouse on the "Leaders" tab at the top of the
page and clicking on the year 2009 under "Batting Leaders," one finds a page
with several choices. Click on the "Fielding" tab, then select shortstops ("SS"
tab), and in the drop-down menu labeled "Min Inn" choose zero (so that every
player who played shortstop is included). If not selected by default, click on
the "Advanced" tab.[5] The website now displays a table featuring defensive
values for 2009 shortstops according to several fielding metrics. On the top-
right angle of the table there is an "Export Data" tag, which allows to save
the table in a comma-delimited (csv) format.

   Once the csv file has been saved, it can easily be read into R. The first
column has some unusual characters in the header of the first column and we
rename the first column to "Name" to remove these characters.

```
fg <- read.csv("FanGraphs Leaderboard.csv")
names(fg)[1] <- "Name"
head(fg)
          Name Team Pos    Inn rSB rGDP rARM rGFP rPM DRS BIZ
1    Jack Wilson - - -  SS  917.1  NA    2   NA    1  25  28 271
2   Elvis Andrus  TEX  SS 1238.0  NA    2   NA    2  11  15 362
3   Paul Janish   CIN  SS  592.1  NA    0   NA    1  10  11 194
4 Cesar Izturis  BAL  SS  934.2  NA    0   NA    3  11  14 312
5 Alex Gonzalez - - -  SS  948.1  NA    0   NA    1  -6  -5 250
6  Brendan Ryan  STL  SS  830.2  NA    2   NA    1  19  22 318
   Plays   RZR OOZ CPP RPP   TZL FSR ARM  DPR RngR ErrR  UZR
```

---

[4]www.fangraphs.com.
[5]The table can be reached directly at tinyurl.com/fgSS09.

```
1   223 0.823  48  NA  NA  8.4  10  NA   1.4 13.9 -0.2 15.1
2   307 0.848  45  NA  NA  6.1  14  NA   0.6 14.1 -2.6 12.1
3   167 0.861  20  NA  NA 11.9   6  NA   0.8  6.5  3.4 10.7
4   259 0.830  39  NA  NA 16.1   8  NA  -0.9  6.5  3.5  9.1
5   204 0.816  31  NA  NA -3.3  -1  NA   1.5  2.4  3.8  7.7
6   266 0.837  54  NA  NA 12.4  11  NA   2.6  1.4  3.3  7.2
  UZR.150 playerid
1    22.7     1017
2    13.5     8709
3    22.7     7412
4    12.1      656
5    12.2      520
6     9.6     6073
```

The `merge` function could be used to combine the *Wizardry*'s and the *FanGraphs*' data frames if they contained a common column. Unfortunately, `wzr` does not contain any player identifiers, thus the most likely candidate columns for the joining of the two datasets are those labeled `Name`. In the following R code, we use the `setdiff` function (which performs asymmetric difference on two sets) to look for unmatched names between the two data frames. The `fg.mismatches` vector contains elements of the column `Name` in the `fg` data frame not found among the elements of the column `Name` in the `wzr` data frame. The `wzr.mismatches` vector will contain differences in the opposite direction.

```
fg.mismatches <- setdiff(fg$Name, wzr$Name)
wzr.mismatches <- setdiff(wzr$Name, fg$Name)


fg.mismatches
[1] "Brent Lillibridge"   "Anderson Hernandez"
[3] "Willie Bloomquist"   "Yuniesky Betancourt"


wzr.mismatches
[1] "A. Hernandez"    "B. Lillibridge" "W. Bloomquist"
[4] "Y. Betancourt"
```

We have learned that Humphreys has elected to use the initial of the first name for some players with a long last name. While in this case, with only four elements on each side, matching by hand would be feasible, but it is helpful to demonstrate a way for identifying similar strings, which could be useful in case of longer lists of elements to match. The `stringdist` package provides functions for computing so-called string distances. The function `stringdistmatrix` compares the elements of the `fg.mismatches` and the `wzr.mismatches` vectors, yielding the distance matrix shown in the following code.

```
library(stringdist)
```

```
dm <- stringdistmatrix(fg.mismatches, wzr.mismatches)
dm
      [,1] [,2] [,3] [,4]
[1,]   15    4   15   16
[2,]    7   16   17   16
[3,]   16   15    5   13
[4,]   16   19   16    7
```

To interpret the above results, the first row of the matrix displays the string distances between the first element of `fg.mismatches` (i.e., "Brent Lillibridge") and every element of `war.mismatches`. The smallest distance has a value of four and is found on the second column of the matrix, corresponding to the second element of `wzr.mismatches` (which is "B. Lillibridge"). Having not specified a value for the `method` argument, the default value `"osa"`, corresponding to the Optimal string alignment, has been used. The method, also known as the Damerau-Levenshtein distance, computes the distance between two strings as the minimum number of operations needed to transform one string into the other. The operations can be either the insertion of a character, the deletion of a character, the substitution of a character, and the transposition of two adjacent characters.

The next step consists of creating a data frame that maps the mismatched names from FanGraphs with the mismatched names from *Wizardry*. We first write a function `index.min` that inputs a vector and returns the index of the element featuring the minimum value. Then the `apply` function is used to execute `index.min` on the first margin (i.e., the rows) of the `dm` matrix. Finally the `names.mapped` data frame is built, matching the mismatched FanGraphs names to the appropriate mismatched Wizardry names.

```
index.min <- function(v) which(v == min(v))
idx <- apply(dm, MARGIN=1, FUN=index.min)
names.mapped <- data.frame(fgName=fg.mismatches,
  wzrName=wzr.mismatches[idx])
```

The last step before merging the two data frames consists of adding a column in one data frame (we choose `wzr`), mapping each name (both those matched and those initially unmatched) to the names of the other data frame. The first line of the following code merges the `wzr` data frame to the `names.mapped` just created. By setting the `all.x` argument to `TRUE`, we request to keep every element of the first data frame passed to `merge` (i.e., `wzr`); the elements of `wzr` that do not have a match in `names.mapped` will feature empty values in the columns of `wzr`. The second line fills those values by copying from the `Name` column whenever the `fgName` column has a missing (`NA`) value.

```
wzr <- merge(wzr, names.mapped, by.x="Name", by.y="wzrName"
  , all.x=TRUE)
wzr$fgName <- ifelse(is.na(wzr$fgName), wzr$Name, wzr$fgName)
```

12.2.6 Applying a function on multiple columns (`plyr`)

After the mapping of names in the previous section, we are now ready to merge the data coming from the two different sources. This is accomplished with the `merge` function, keeping only the columns needed for the further analysis.

```
defense <- merge(fg, wzr[,-1], by.x="Name", by.y="fgName")
defense <- defense[,c("Name", "Inn", "Plays", "UZR", "DRS"
 , "TZL",  "RZR", "FSR", "Runs")]
```

Some of the players have played very little at the shortstop position in 2009 and thus may not have been rated by one or more of the selected defensive metrics.[6] One can choose to ignore them when computing pairwise correlations between two metrics.[7] Another option is to assume those players with limited playing time to have performed at an average rate, thus substituting the missing values with zeros for the defensive metrics expressed in runs.[8]

A new function `coalesce` is written that takes a value `x` as input and returns another value `dflt` (zero if not otherwise specified) whenever `x` is NA. In the lines of code that follow we apply `coalesce` to the RZR column, substituting NAs with the average value of .801.

```
coalesce <- function(x, dflt=0) ifelse(is.na(x), dflt, x)
defense$RZR <- coalesce(defense$RZR, dflt=.801)
```

The `coalesce` function is then applied to every column of `defense`. One can avoid calling this function multiple times thanks to the `colwise` function in the `plyr` package which takes a vector-input function as the argument and returns another function which works columnwise on a data frame.

```
library(plyr)
coalesceColumns <- colwise(coalesce)
defense <- coalesceColumns(defense)
```

12.2.7 Weighted correlations (`weights`)

The function `cor` computes pairwise correlations between the various defensive metrics contained in the data frame `defense`. The `[,-1:-3]` indexing excludes the first three columns (the player's name, the innings he played and the plays he made) from the computations. The `round` function is used to display only three decimal digits.

---

[6]Try typing the command `subset(defense, Inn < 5)` in the R console for displaying some of them.

[7]Such exclusion would be obtained by specifying the value `"pairwise.complete.obs"` to the `use` argument of the `cor` function.

[8]Since RZR is expressed as a rate of succesful plays, we will substitute NAs with the average success rate, which was .801 for shortstops in 2009.

```
round(cor(defense[,-1:-3]),3)
       UZR   DRS   TZL   RZR   FSR  Runs
UZR  1.000 0.782 0.727 0.142 0.489 0.305
DRS  0.782 1.000 0.724 0.090 0.596 0.387
TZL  0.727 0.724 1.000 0.124 0.495 0.322
RZR  0.142 0.090 0.124 1.000 0.050 0.059
FSR  0.489 0.596 0.495 0.050 1.000 0.089
Runs 0.305 0.387 0.322 0.059 0.089 1.000
```

Every observation (i.e., every row in the `defense` data frame) counts the same when calculating the correlations. This may be inappropriate – a shortstop who has more opportunities perhaps should have greater weight than a part-time shortstop in the computation of the correlations. The function `wtd.cor` in the `weights` package can compute a correlation allowing for different weights for the observations. The function `wtd.cor` has two inputs: the matrix of observations and a vector of weights which in this case corresponds to the number of innings played. The function returns three matrices: a matrix of correlation coefficients, and matrices of t-values and p-values. In the code that follows we limit the output to the correlation matrix, which can be compared to the one produced by `cor`.

```
library(weights)
round(wtd.cor(defense[,-1:-3], weight=defense$Inn)$correlation, 3)
       UZR   DRS   TZL   RZR   FSR  Runs
UZR  1.000 0.794 0.731 0.618 0.502 0.241
DRS  0.794 1.000 0.745 0.433 0.598 0.364
TZL  0.731 0.745 1.000 0.565 0.487 0.288
RZR  0.618 0.433 0.565 1.000 0.275 0.171
FSR  0.502 0.598 0.487 0.275 1.000 0.018
Runs 0.241 0.364 0.288 0.171 0.018 1.000
```

The matrix of weighted correlations is slightly different than the raw (unweighted) correlations. In particular, it appears the very low correlation between the Revised Zone Rating (`RZR`) and the other metrics was highly affected by the numbers posted by the occasional players.

12.2.8   Displaying correlation matrices (`ellipse`)

A correlation matrix can be displayed graphically by the use of ellipse-shaped glyphs (Murdoch and Chow (1996)). The ellipses will have a bottom-left to top-right orientation for positive correlations and a bottom-right to top-left orientation for negative correlations. They will appear narrower for higher levels of correlations and will look more like circles as the correlation coefficients approximate zero.

The package `ellipse` provides the function `plotcorr` to produce such visualizations. The results of the following code can be seen in Figure 12.1

```
library(ellipse)
Dcor <- wtd.cor(defense[,-1:-3], weight=defense$Inn)$correlation
plotcorr(Dcor)
```

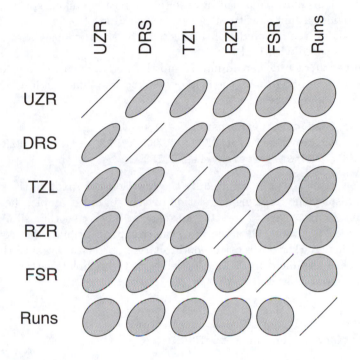

FIGURE 12.1
Visualization of the correlation matrix for defensive metrics.

12.2.9   Evaluating the fielding metrics (psych)

The ellipses in Figure 12.1 indicate positive correlations between any pair of defensive metrics, but the strengths of the associations among the various fielding metrics are variable, as shown graphically by the thickness of the ellipses and numerically in the correlation matrix in Section 12.2.7. Not surprisingly, UZR and DRS have the highest correlation (0.79) – this is understandable since both measures are based on Baseball Info Solution data. The measure FSR, based solely on subjective evaluations, shows a fair degree of association with some of the other metrics. On the other hand, the method outlined in *Wizardry*, labeled as *Runs*, displays the lowest correlations across the board.

By following steps similar to the ones outlined in Sections from 12.2.3 to 12.2.5, we created a file named `teamD.csv`, which contains team cumulative defensive ratings for the seasons 2007 to 2009. In addition to the six metrics previously presented, this file features an additional rating, the Defensive Efficiency Record, labeled DER. This fielding measure, first proposed by Bill James, is calculated by dividing the outs recorded on batted balls by the number of balls playable by the defense (every batted ball except home runs). DER provides a fair, albeit not perfect, estimation of defensive value at the team level.[9] In the remainder of this section the six previously shown metrics will be compared with DER.

The teamD.csv file is read into R and the weighted correlations of the six metrics are computed, with the help of the wtd.cor function in the weights package, as done in Section 12.2.7.

```
teamD <- read.csv("teamD.csv")
Dcor <- wtd.cor(teamD[,c("DER", "UZR", "DRS", "TZL", "RZR"
  , "FSR", "Runs")], weight=teamD$IP)
```

A new visualization, produced by the function cor.plot in the psych, is used to display the resulting correlation matrix. In our example, the function mat.sort is first used to sort the correlation matrix so that similar items are grouped together. Then the function cor.plot displays the correlation matrix where darker boxes correspond to larger correlation values. By setting the numbers argument to TRUE), the values of the correlations, on a 0 to 100 scale, are also displayed.

```
library(psych)
sortedCor <- mat.sort(Dcor$correlation)
cor.plot(sortedCor, numbers=TRUE)
```

According to Figure 12.2, TZL has the highest correlation with DER for the three-year period.

12.3   Comparing Two Shortstops

In this section, the long careers[10] of two great shortstops are used to illustrate functions available in several contributed R packages. Specifically, suppose one is interested in visualizing fielding ratings of the careers of Yankee captain Derek Jeter and long-time Phillies player Jimmy Rollins according to the metrics UZR and DSR. The FanGraphs website allows registered users[11] to

---

[9]The main shortcoming of DER consists in valuing all unsuccessful defensive plays the same, thus not accounting for the different outcomes of batted balls (singles, doubles, and so on).

[10]We will actually be using part of the career of Rollins and Jeter, particularly the seasons covered by BIS data, thus having UZR and DRS values available.

[11]Registration is free of charge.

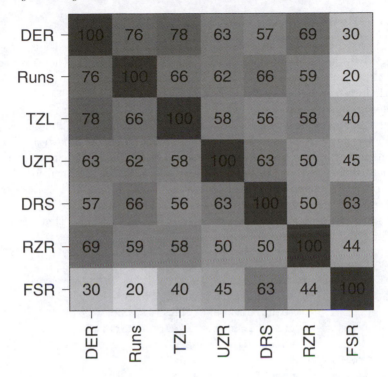

FIGURE 12.2
Correlation plot for the comparison of team defensive metrics.

create custom leaderboards, by selecting desired stats and a subset of players. For this particular example, we request a table displaying Jeter's and Rollins' year-by-year UZR and DRS measures and the resulting table is exported in the `jeter_rollins.csv` file. In the following R code, the file is loaded into the `jetrol` data frame and its first few lines are displayed.

```
jetrol <- read.csv("jeter_rollins.csv")
head(jetrol)
  Season         Name Team DRS  UZR playerid
1   2008 Jimmy Rollins  PHI  18 12.9      971
2   2003 Jimmy Rollins  PHI  11  7.4      971
3   2010 Jimmy Rollins  PHI   3  6.9      971
4   2004 Jimmy Rollins  PHI   9  6.8      971
5   2009   Derek Jeter  NYY   3  6.4      826
6   2006 Jimmy Rollins  PHI  12  5.0      971
```

12.3.1 Reshaping the data (reshape2)

Data in the `jetrol` data frame require a little reshaping in order to be easily processed through one of the powerful graphics packages (either `lattice` or `ggplot2`). The function `melt` in the `reshape2` package is used to give the `jetrol` data frame the desired shape.

The arguments of the function `melt` are the data frame, `id.vars`, `measure.vars`, `variable.name`, and `value.names`. In this example, we indicate, by the `id.vars` argument, that the new data frame will have the first two columns featuring the season and the name of the player. The `measure.vars` argument indicates that the measured variables `DRS` and `UZR` will be the ones undergoing the reshaping – two new columns will appear in the data frame. The `variable.name` argument indicates that `fieldingMetric` is the name of the defensive metric, and the `value.name` argument indicates that `runs` is the name of the defensive rating.

```
library(reshape2)
jetrol2 <- melt(jetrol
                , id.vars=c("Season", "Name")
                , measure.vars=c("DRS", "UZR")
                , variable.name="fieldingMetric"
                , value.name="runs")
head(jetrol2)
  Season          Name fieldingMetric runs
1   2008 Jimmy Rollins            DRS   18
2   2003 Jimmy Rollins            DRS   11
3   2010 Jimmy Rollins            DRS    3
4   2004 Jimmy Rollins            DRS    9
5   2009   Derek Jeter            DRS    3
6   2006 Jimmy Rollins            DRS   12
```

12.3.2 Plotting the data (ggplot2 and directlabels)

The `jetrol2` data frame is now ready to be plotted using the `ggplot2` package, extensively introduced in Chapter 6. The `ggplot` object is initialized by mapping the `Season` to the x-axis, the `runs` to the y-axis, and the `fieldingMetrics` to the color aesthetic. By adding the `geom_line` object, we indicate we want a line graph, and the `facet_grid` function is used to place the players (`Name`) in separate panels. The `scale_color_manual` layer is used for manually selecting the line colors (the label appearing as the legend title is specified here as well), while in the `scale_x_continuous` layer the position of the tick marks is indicated on the x-axis. Finally a dotted reference line is drawn at zero runs saved. Since the whole `ggplot` object has been assigned to the variable `p`, one needs to type `p` in the R console to actually display the plot.

```
library(ggplot2)
p <- ggplot(jetrol2, aes(x=Season, y=runs
  , col=fieldingMetric)) +
      geom_line() +
      facet_grid(. ~ Name) +
      scale_color_manual(name="Fielding\nmetric"
        , values=c("black", "grey70")) +
      scale_x_continuous(breaks=seq(2004, 2012, 4)) +
      geom_hline(yintercept=0, lty=3)
```

The package `directlabels` provides a quick-to-use function `direct.label` which adds labels directly to the plot, thus removing the color legend, as shown in Figure 12.3.[12]

```
library(directlabels)
direct.label(p)
```

The plots in Figure 12.3 show that Jeter has been constantly rated as a below-average fielder (despite both UZR and DRS indicating an unusually good season for him in 2009), while Rollins has mostly been an above-average fielding shortstop. UZR and DRS numbers follow more or less the same path for the Yankee captain, but their portrait of the Philadelphia shortstop convey different information: where DRS seems to indicate a steep decline in Jimmy's defensive value, he appears to have maintained his ability according to UZR.

12.4 Further Reading

Rickey (1954) is an article that appeared on the August 2, 1954, issue of *Life* magazine, in which Branch Rickey, considered one of the greatest baseball executives ever, illustrates what are among the earliest known sabermetric-like analyses. On page 83 of the magazine, he admits that fielding value as was measured at the time (with fielding percentage) is "utterly worthless." Humphreys (2011) outlines how several of the fielding metrics developed in the past 15 years of work, and illustrates a method of his own to estimate runs saved by fielders using data available throughout most of baseball history. John Dewan (2006, 2009, 2012) periodically publishes a volume of The Fielding Bible, containing essays and fielding leaderboard tables based on the Baseball Info Solution proprietary system.

---

[12]Note that in most cases, adding labels to the area where data are displayed unnecessarily clutters the plot.

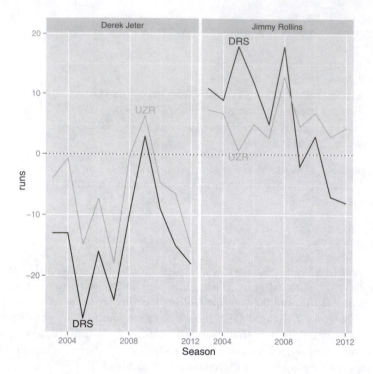

FIGURE 12.3
Derek Jeter's and Jimmy Rollins' defensive value through the years as measured by UZR and DRS. Labels added directly on the plot area.

12.5   Exercises

1. **(Data Reshaping: Exploring Team Plate Discipline)**

    (a) On the FanGraphs website (`www.fangraphs.com/`) generate a report containing team batting plate discipline stats from 2008 to the latest completed season. Export the data as a csv file and load it into R. Rename the columns if necessary.

    (b) Calculate the correlation between percentage of swings outside the zone (`O-Swing%`) and the percentage of pitches in the strike zone (`Zone%`)). Are teams with a high propensity toward swinging at bad pitches fed with a lower number of pitches in the strike zone?

    (c) Reshape the loaded data so that the new data frame will consist of four columns: season, team, the name of the discipline statistic, and the value of the discipline statistic.

(d) Subset the data frame so that it only contains O-Swing% and Zone% rows.

(e) Draw a plot where teams are displayed in different panels, each featuring the year-to-year trend of both O-Swing% and Zone%.

2. (**Applying a Function on Multiple Columns: Exploring the Availability of Statistics Through Baseball History**)

Not every baseball statistic has been tracked since the early days of baseball. Stolen Bases and Caught Stealing are examples of numbers which are not available in some earlier seasons.

(a) Load the batting statistics from Lahman's database and create a new data frame containing only the following information: year, league, stolen bases, and runners caught stealing.

(b) Write a function that takes a vector as its only argument and returns the percentage of missing values (NAs) in the vector. Note that is.na is a function for identifying NAs and that the function length returns the number of elements in a vector.

(c) Apply the newly written function to the columns of the previously created data frame.

(d) Use the ddply function (from the plyr package) to apply the function by year and by league. Was the recording of SBs and CSs introduced at the same time in the National and the American League?

3. (**Comparing Defensive Ratings and Stealing Ability in Centerfielders**)

As speed is an important trait of a centerfielder skill set, good base stealers are often found among players manning the middle outfielder position.

(a) Download seasonal fielding ratings for centerfielders from *Wizardry*'s online resources. Read the Excel file into R.

(b) Using tables from Lahman's database, prepare a data frame containing seasonal data on offensive stolen bases and caught stealing for players who have appearances (in the given season) at centerfield.

(c) Estimates like those performed in Chapter 5 indicate that successful stolen bases are worth roughly 0.2 runs to the offensive team, while a caught stealing cost it 0.5 runs. Using these values compute stolen base runs for every player-season.

(d) Merge defensive ratings with stolen base rating data. Note that on the previous point you may want to gather the players' first and last names from the Master table. Try to perform string matching as shown in Section 12.2.5 to match the highest number of players-seasons.

(e) Compute the correlation between Stolen Base Runs and Defensive Runs (as reported in *Wizardry*).

A

Retrosheet Files Reference

CONTENTS

A.1   Downloading Play-by-Play Files 301
     A.1.1   Introduction 301
     A.1.2   Setup 302
     A.1.3   Using a special function for a particular season 302
     A.1.4   Reading the files into R 302
     A.1.5   The function `parse.retrosheet.pbp` 302
A.2   Retrosheet Event Files: a Short Reference 304
     A.2.1   Game and event identifiers 304
     A.2.2   The state of the game 305
A.3   Parsing Retrosheet Pitch Sequences 306
     A.3.1   Introduction 306
     A.3.2   Setup 306
     A.3.3   Evaluating every count 307

A.1   Downloading Play-by-Play Files

A.1.1   Introduction

The play-by-play data files are currently available at the Retrosheet Web page `www.retrosheet.org/game.htm`. By clicking on a single year, say 1950, one obtains a compressed (zip) file containing a collection of files, one set of files containing information on the plays for the home games for all teams, and another set of files giving the rosters of the players for each team. This appendix describes the use of a single function `parse.retrosheet.pbp` to create a single file, in csv format, of all plays for games played in a particular season, and also create a csv file of the rosters of all players. We assume that one is working in a Windows session, although similar operations can be performed on other systems.

A.1.2   Setup

In the current R working directory, create a new folder "download.folder" and in this new folder, create two folders "zipped" and "unzipped." There are special software tools for working with the Retrosheet files that should be downloaded from `sourceforge.net/projects/chadwick/files/`. When the compressed file is expanded, the file "cwevent.exe" should be placed in the "unzipped" folder.

A.1.3   Using a special function for a particular season

Open R and make sure that you are in the working directory containing the folder "download.folder." Read in the function "parse.retrosheet.pbp.R" by use of the `source` function.

```
source("parse.retrosheet.pbp.R")
```

The function `parse.retrosheet.pbp` has a single argument `season`. To obtain the play-by-play files for the 1950 season, the relevant Retrosheet files are obtained by typing

```
parse.retrosheet.pbp(1950)
```

If you look in the "download.folder/unzipped" folder, you will see two new files: "all1950.csv" contains the play for play records for the 1950 season, and "roster1950.csv" contains roster information for all players of that season.

A.1.4   Reading the files into R

To read the data files into R, one should first make sure that the files "all1950.csv" and "roster1950.csv" are in the current working directory. In addition, the file "fields.csv" containing the names of all of the fields in the play-by-play record file should be placed in this directory. This file was created by the authors from the information in the page `chadwick.sourceforge.net/doc/cwevent.html`. The following R script will read in the three files by three applications of the `read.csv` function. In addition, the `names` function is used to assign names to the play-by-play data frame using the `Header` variable from the `fields` data frame.

```
data <- read.csv("all1950.csv", header=FALSE)
roster <- read.csv("roster1950.csv")
fields <- read.csv("fields.csv")
names(data) <- fields[, "Header"]
```

A.1.5   The function `parse.retrosheet.pbp`

The R function `parse.retrosheet.pbp` listed below contains five functions. We briefly describe what is done in each of these functions.

The function `download.retrosheet` downloads the Retrosheet zip file from the particular season and saves the zip file in the "zipped" folder. The function `unzip.retrosheet` unzips the file and puts the individual event files and roster files in the "unzipped" folder. The function `create.csv.file(season)` reads the individual team record files into R, merges the files, and writes a new single combined file in csv format in the "unzipped" folder. The function `create.csv.roster` function does a similar thing with the individual roster files. Last, the `cleanup` function removes the individual record and roster files that are no longer needed.

```
parse.retrosheet.pbp <- function(season){
download.retrosheet <- function(season){
  download.file(
    url=paste("http://www.retrosheet.org/events/", season, "eve.zip", sep="")
    , destfile=paste("download.folder", "/zipped/", season, "eve.zip", sep="")
    )
}
unzip.retrosheet <- function(season){
  unzip(paste("download.folder", "/zipped/", season, "eve.zip", sep=""),
        exdir=paste("download.folder", "/unzipped", sep=""))
}
create.csv.file <- function(year){
  wd <- getwd()
  setwd("download.folder/unzipped")
  shell(paste(paste("cwevent -y", year, "-f 0-96"),
              paste(year, "*.EV*", sep=""),
              paste("> all", year, ".csv", sep="")))
  setwd(wd)
}
create.csv.roster <- function(year){
  filenames <- list.files(path = "download.folder/unzipped/")
  filenames.roster <-
    subset(filenames, substr(filenames, 4, 11) == paste(year,".ROS",sep=""))
  read.csv2 <- function(file)
    read.csv(paste("download.folder/unzipped/", file, sep=""), header=FALSE)
  R <- do.call("rbind", lapply(filenames.roster, read.csv2))
  names(R)[1 : 6] <- c("Player.ID", "Last.Name", "First.Name",
                       "Bats", "Pitches", "Team")
  wd <- getwd()
  setwd("download.folder/unzipped")
  write.csv(R, file=paste("roster", year, ".csv", sep=""))
  setwd(wd)
}
cleanup <- function(){
  wd <- getwd()
  setwd("download.folder/unzipped")
  shell("del *.EVN")
  shell("del *.EVA")
  shell("del *.ROS")
```

```
  shell("del TEAM*")
  setwd(wd)
}
  download.retrosheet(season)
  unzip.retrosheet(season)
  create.csv.file(season)
  create.csv.roster(season)
  cleanup()
}
```

A.2   Retrosheet Event Files: a Short Reference

As was mentioned in Chapter 1, Retrosheet event files come in a format expressly devised for them, and require the use of some software tools for converting them in a format suitable for data analysis. Retrosheet provides such software tools (`www.retrosheet.org/tools.htm`) and a step-by-step example (`www.retrosheet.org/stepex.txt`) for performing the conversion.

Another website[1] (whose name is a tribute to baseball pioneer Henry Chadwick credited for devising the box score) provides similar tools for parsing Retrosheet event files,[2] that have been used for creating the play-by-play files used in this book.

Chadwick tools generate a line for each play in the Retrosheet event files, consisting of 97 "regular" columns (the same that are obtained using the tools provided by Retrosheet) plus 63 "extended" fields, allowing to easily access all of the information contained in the Retrosheet event files. Going through every one of the more than 150 columns generated by the Chadwick tools is beyond the scope of this book, thus we point to the documentation on the Chadwick website for the full list.[3] In this section we present the main fields describing an event and the state of the game when it happens.

A.2.1   Game and event identifiers

The games are identified in Retrosheet event files by 12-character strings (the `GAME_ID` column): the first three characters identify the home team, the following eight characters indicate the date when the game took place (in the YYYYMMDD format), and the last character is used to distinguish games of

---

[1] *Chadwick: Software Tools for Scoring Baseball Games* `chadwick.sourceforge.net/doc/index.html`.

[2] Download link: `sourceforge.net/projects/chadwick/files`.

[3] The documentation for all the software tools is available at `chadwick.sourceforge.net/doc/cwtools.html`. In particular, the tool for processing the event files (*cwevent*) is documented at `chadwick.sourceforge.net/doc/cwevent.html#cwtools-cwevent`.

doubleheaders (thus "1" indicates the first game, "2" the second game, and "0" means only one game was played on the day).

Events are progressively numerated in each game (column EVENT_ID), thus every single action in the Retrosheet database can be uniquely identified by the combination of the game identifier and the event identifier.

A.2.2 The state of the game

Several fields are helpful for defining the state of the game when a particular event happened. The **inning** and the **team on offense** variables are stored in the INN_CT and BAT_HOME_ID fields respectively. The latter field can assume values "0" (away team batting, i.e., top of the inning) or "1" (home team batting, bottom of the inning). The **visitor score** and the **home score** variables are recorded in the AWAY_SCORE_CT and HOME_SCORE_CT.

The **number of outs** before the play is indicated in the OUTS_CT column, while the situation of `runners on base` is coded in the field START_BASES_CD, using numbers from 1 to 7 as shown in Table A.1.[4]

TABLE A.1
Retrosheet coding for the situation of runners on base.

Code	Bases occupancy
0	Empty
1	1B only
2	2B only
3	1B & 2B
4	3B only
5	1B & 3B
6	2B & 3B
7	Loaded

The actual **description of the event** resides in the EVENT_TX column, consisting in a string describing the outcome of the play (e.g., strikeout, single, etc.), some additional details (e.g. the type and location of the batted ball), and the advancement of any runner on base. Several columns are generated by decoding the EVENT_TX string:

- EVENT_CD is a numeric code reflecting the **basic event**; Table A.2 displays the codes for the possible plays coded in this column.

- BAT_EVENT_FL is a flag indicating whether an event is a **batting event**, in which case it is labeled as T. Non-batting events include, for example,

---

[4]An analogous column named END_BASES_CD contains the base state at the end of the play, coded in the same way.

stolen bases, wild pitches and, generally, any event that does not mark the end of a plate appearance.

- H_CD is a numeric code indicating the **base hit type**, going from 1 for a single to 4 for a home run.

- BATTEDBALL_CD is a single character code denoting the **batted ball type**. It can assume one of the following values: G (ground ball), L (line drive), F (fly ball), P (pop-up). Note that for most of the seasons in the Retrosheet database, the batted ball type is reported only for plate appearances ending with the batter making an out, while they are not available on base hits.

- BATTEDBALL_LOC_TX is a string indicating the **batted ball location**, coded according to the diagram shown at www.retrosheet.org/location.htm. Note that this information is available for a limited number of seasons.

- FLD_CD is a numeric code denoting the **fielder** first touching a batted ball, coded with the conventional baseball fielding notation going from 1 (the pitcher) to 9 (the right fielder).

The **sequence of pitches** is recorded in the PITCH_SEQ_TX and has been addressed in Chapter 7, where Table 7.1 displays how the different pitch outcomes are coded. Several columns are generated from this one, indicating counts of the various types of pitch outcomes, as displayed in Table A.3.

A.3   Parsing Retrosheet Pitch Sequences

A.3.1   Introduction

Chapter 7 showed how to compute, by the use of regular expressions, whether a plate appearance went through either a 1-0 or a 0-1 count. Here the code is provided to retrieve the same information for every possible balls/strikes count.

A.3.2   Setup

We first load Retrosheet data for the 2011 season.

```
pbp2011 <- read.csv("retrosheet/all2011.csv")
headers <- read.csv("retrosheet/fields.csv")
names(pbp2011) <- headers$Header
```

TABLE A.2
Retrosheet coding for the type of event.

Code	Event type
2	Generic Out
3	Strikeout
4	Stolen Base
5	Defensive Indifferen
6	Caught Stealing
8	Pickoff
9	Wild Pitch
10	Passed Ball
11	Balk
12	Other Advance
13	Foul Error
14	Nonintentional Walk
15	Intentional Walk
16	Hit By Pitch
17	Interference
18	Error
19	Fielder Choice
20	Single
21	Double
22	Triple
23	Homerun

Then a new column `sequence` is created in which the pitch sequence is reported, stripped by any character not indicating an actual pitch to the batter.[5]

```
pbp2011$sequence <- gsub("[.>123+*N]", "", pbp2011$PITCH_SEQ_TX)
```

A.3.3 Evaluating every count

Every plate appearance starts with a 0-0 count. The code for both the 1-0 and 0-1 counts was described in Chapter 7.

```
pbp2011$c00 <- TRUE
pbp2011$c10 <- grepl("^[BIPV]", pbp2011$sequence)
pbp2011$c01 <- grepl("^[CFKLMOQRST]", pbp2011$sequence)
```

A number inside the square brackets indicates the minimum number of times the preceding expression has to be repeated in the string to be parsed. The following lines look for plate appearances going through the counts 2-0, 3-0, and 0-2.

---

[5]See Table 7.1 in Chapter 7 for reference.

TABLE A.3
Columns reporting counts of various pitch types.

Column name	Column description
PA_BALL_CT	No. of balls in plate appearance
PA_CALLED_BALL_CT	No. of called balls in plate appearance
PA_INTENT_BALL_CT	No. of intentional balls in plate appearance
PA_PITCHOUT_BALL_CT	No. of pitchouts in plate appearance
PA_HITBATTER_BALL_CT	No. of pitches hitting batter in plate appearance
PA_OTHER_BALL_CT	No. of other balls in plate appearance
PA_STRIKE_CT	No. of strikes in plate appearance
PA_CALLED_STRIKE_CT	No. of called strikes in plate appearance
PA_SWINGMISS_STRIKE_CT	No. of swinging strikes in plate appearance
PA_FOUL_STRIKE_CT	No. of foul balls in plate appearance
PA_INPLAY_STRIKE_CT	No. of balls in play in plate appearance
PA_OTHER_STRIKE_CT	No. of other strikes in plate appearance

```
pbp2011$c20 <- grepl("^[BIPV]{2}", pbp2011$sequence)
pbp2011$c30 <- grepl("^[BIPV]{3}", pbp2011$sequence)
pbp2011$c02 <- grepl("^[CFKLMOQRST]{2}", pbp2011$sequence)
```

The | (vertical bar) character is used to separate alternatives. The following lines parse the sequence string looking for the different sequences that can lead to 1-1, 2-1, and 3-1 counts.

```
pbp2011$c11 <- grepl("^([CFKLMOQRST][BIPV]|[BIPV][CFKLMOQRST])"
  , pbp2011$sequence)
pbp2011$c21 <- grepl("^([CFKLMOQRST][BIPV][BIPV]
  |[BIPV][CFKLMOQRST][BIPV]
  |[BIPV][BIPV][CFKLMOQRST])", pbp2011$sequence)
pbp2011$c31 <- grepl("^([CFKLMOQRST][BIPV][BIPV][BIPV]
  |[BIPV][CFKLMOQRST][BIPV][BIPV]|[BIPV][BIPV][CFKLMOQRST][BIPV]
  |[BIPV][BIPV][BIPV][CFKLMOQRST])", pbp2011$sequence)
```

On two-strike counts, batters can indefinitely foul pitches off without affecting the count. In the lines below, sequences reaching two strikes before reaching the desired number of balls feature the [FR]* expression, denoting a foulball[6] happening any number of times, including zero, as indicated by the asterisk.

```
pbp2011$c12 <- grepl("^([CFKLMOQRST][CFKLMOQRST][FR]*[BIPV]
  |[BIPV][CFKLMOQRST][CFKLMOQRST]
  |[CFKLMOQRST][BIPV][CFKLMOQRST])", pbp2011$sequence)
```

[6]F encodes a foulball, R a foulball on a pitchout. See Table 7.1.

```
pbp2011$c22 <- grepl("^(
                     [CFKLMOQRST] [CFKLMOQRST] [FR]*[BIPV] [FR]*[BIPV]
                     | [BIPV] [BIPV] [CFKLMOQRST] [CFKLMOQRST]
                     | [BIPV] [CFKLMOQRST] [BIPV] [CFKLMOQRST]
                     | [BIPV] [CFKLMOQRST] [CFKLMOQRST] [FR]*[BIPV]
                     | [CFKLMOQRST] [BIPV] [CFKLMOQRST] [FR]*[BIPV]
                     | [CFKLMOQRST] [BIPV] [BIPV] [CFKLMOQRST]
                     )", pbp2011$sequence)
pbp2011$c32 <- grepl("^[CFKLMOQRST]*[BIPV] [CFKLMOQRST]*
  [BIPV] [CFKLMOQRST]*[BIPV]", pbp2011$sequence)
  & grepl("^[BIPV]*[CFKLMOQRST] [BIPV]*[CFKLMOQRST]"
  , pbp2011$sequence)
```

B

Accessing and Using MLBAM Gameday and PITCHf/x Data

CONTENTS

B.1   Introduction 311
B.2   Where are the Data Stored? 312
B.3   Suitable Formats for PITCHf/x Data 314
     B.3.1   Obtaining data from on-line resources 314
     B.3.2   Parsing in R 314
         B.3.2.1   A wrapper function 315
B.4   Details on the Data 316
     B.4.1   `atbat` attributes 316
     B.4.2   `pitch` attributes 317
     B.4.3   `hip` attributes (hit locations data) 318
B.5   Special Notes About the Gameday and PITCHf/x Data 319
B.6   Miscellanea 320
     B.6.1   Calculating the pitch trajectory 320
     B.6.2   An R package for getting and visualizing PITCHf/x
         data: `pitchRx` 321
     B.6.3   Cross-referencing with other data sources 323
     B.6.4   Online resources 323

B.1   Introduction

This section provides further details on Gameday and PITCHf/x data. It will be shown where those data are stored, how they can be retrieved using resources available on the Web, and how they can be parsed using R. A description of the most important fields will be provided, together with an overview of the issues a researcher should be aware of when analyzing these data.

B.2   Where are the Data Stored?

Data relevant to games of one particular day can be found following a link such
as gd2.mlb.com/components/game/mlb/year_2012/month_06/day_13/. By
simply modifying the year, the month, and the day, one can reach any MLB
game of the Gameday era.[1] Figure B.1 displays the page featuring the games
played on June 13, 2012.

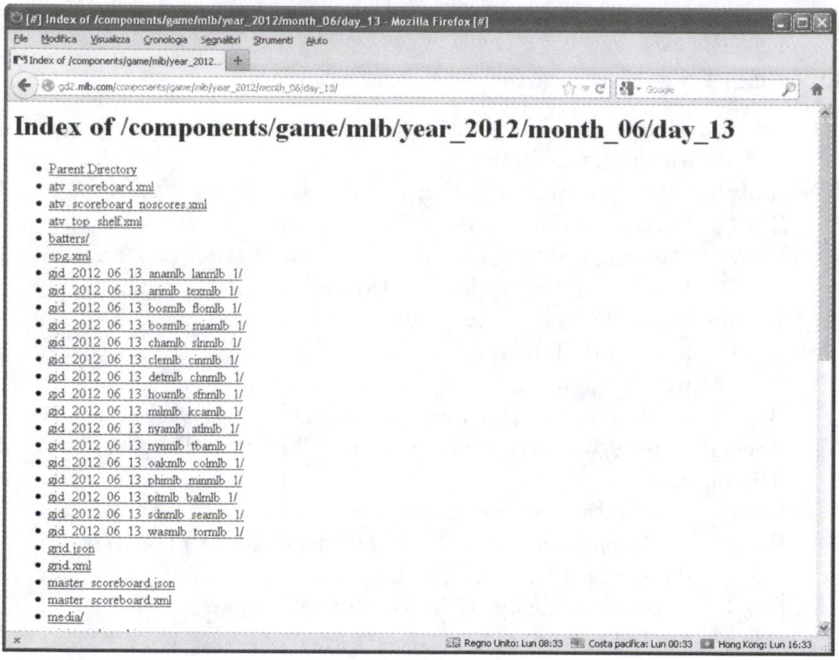

FIGURE B.1
MLB directory of games played on June 13, 2012.

To access data relative to a specific game, one follows one of the
links beginning with the characters gid_. For example, clicking on the
gid_2012_06_13_houmlb_sfnmlb_1/ link, one reaches the page for the Houston
Astros @ San Francisco Giants game, in which Matt Cain pitched a perfect
game. Once the gd2.mlb.com/components/game/mlb/year_2012/month_06/
day_13/gid_2012_06_13_houmlb_sfnmlb_1/ page is reached, all of the play-
by-play data, including PITCHf/x and hit locations, are available by following the inning/ link. PITCHf/x data for the whole game are stored in the

---

[1]Actually, by modifying the mlb part of the URL, one can even access minor league
games.

`inning_all.xml` XML page and they appear on the Web page as in Figure B.2. Similarly, data on hit locations are stored in the `inning_hit.xml` page.

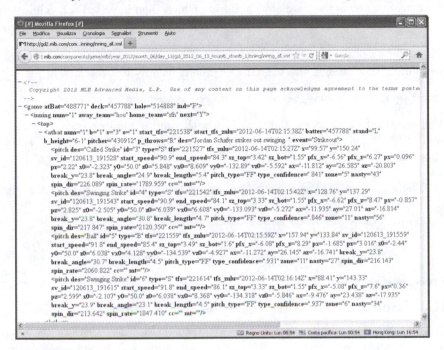

FIGURE B.2
Pitch-by-pitch data of the Astros – Giants game played on June 13, 2012.

The form of XML documents is a tree structure. The `inning_all.xml` files have the following structure.

```
<game>
  <inning>
    <top>
      <atbat>
        <pitch>
      </atbat>
    </top>
    <bottom>
      <atbat>
        <pitch>
      </atbat>
    </bottom>
  </inning>
</game>
```

The `<game>` element is said to be the *root* element of the tree – this indicates that this document is a game. The `<inning>` element is a *child* of the root, and `<top>` and `<bottom>` are in turn children of `<inning>`. Both `<top>` and `<bottom>` elements have `<atbat>` children, which then have `<pitch>` children. Note that all elements are delimited by an `<element>` tag at the beginning and an `</element>` tag at the end.

In the opening tag (the one without the slash) of elements, *attributes* can be found. For example, the `<inning>` element usually has the format

```
<inning num="1" away_team="hou" home_team="sfn" next="Y">
```

The `<inning>` element has four attributes: the inning number (`num` which has value `"1"` in this case), the teams involved (`away_team` and `home_team`, set as `"hou"` for Houston and `"sfn"` for San Francisco), and whether another inning has been played (`next`, which has value `"Y"` in this case).

The attributes of the `<atbat>` and the `<pitch>` elements are the ones containing the play-by-play and pitch-by-pitch information, including PITCHf/x data, and will be described in Sections B.4.1 and B.4.2.

B.3   Suitable Formats for PITCHf/x Data

To analyze pitch-by-pitch data, the XML information needs to be converted in a more suitable format. In this section a few online resources for converting PITCHf/x data to a user-ready form will be described and some details on how to parse XML data in R will be provided.

B.3.1   Obtaining data from on-line resources

Starting from Joseph Adler's publication of *Baseball Hacks* (Adler, 2006), several people have shared their codes to download and parse Gameday data. Kyle Wilkomm developed his parser, named *BBOS* (short for Baseball On a Stick), and made it available at `sourceforge.net/projects/baseballonastic/`. BBOS uses Python code to create and populate a MySQL database of pitch-by-pitch data and can be run by users not familiar with Python. Jeff Zimmerman has further simplified life for aspiring baseball analysts, as he makes his MySQL pitch-by-pitch database (updated daily during the season) available for download at `www.baseballheatmaps.com/pitch-fx-download/` (see Section 11.6.2 for details).

B.3.2   Parsing in R

The `XML` package provides XML parsing functionality in R. The first step in the parsing process reads the XML file into R using the `xmlParse` function.

```
library(XML)
gameUrl <- "gd2.mlb.com/components/game/mlb/year_2012/mo
    nth_06/day_13/gid_2012_06_13_houmlb_sfnmlb_1/inning/inning_al
    l.xml"
xmlGame <- xmlParse(gameUrl)
```

Several functions are available to access specific nodes of the tree. For example, one can obtain the `inning` nodes by using the `getNodeSet` function.

```
xmlInnings <- getNodeSet(xmlGame, "//inning")
length(xmlInnings)
[1] 9
```

By using the `xmlAttrs` function, one can retrieve the attributes of a node. The following code obtains the attributes for the first inning.

```
xmlAttrs(xmlInnings[[1]])
```

```
   num away_team home_team        next
   "1"      "hou"     "sfn"         "Y"
```

To obtain the attributes for all the nine innings, one can use the function `ldply` from the `plyr` package analogous to the function `ddply` introduced in Chapter 2.[2]

```
library(plyr)
ldply(xmlInnings, xmlAttrs)
```

```
  num away_team home_team next
1   1       hou       sfn    Y
2   2       hou       sfn    Y
3   3       hou       sfn    Y
4   4       hou       sfn    Y
5   5       hou       sfn    Y
6   6       hou       sfn    Y
7   7       hou       sfn    Y
8   8       hou       sfn    Y
9   9       hou       sfn    N
```

B.3.2.1 A wrapper function

The few lines presented in the previous section are most of what is needed to parse XML data into R data frames. However, since the corresponding XML elements may not always have the same number of columns (for example, if one pitch did not get tracked by the PITCHf/x system, it will lack all the relevant

---

[2]`ldply` differs from `ddply` only for the input data it accepts: where `ddply` takes a data frame (hence the "d" at the beginning of its name), `ldply` requires a list (hence the "l").

attributes), the `ldply` function might encounter problems in generating the data frame due to the mismatch of attributes. The following wrapper function `grabXML` takes care of this issue and can be used for the parsing of most XML documents.

```
grabXML <- function(XML.parsed, field){
  parse.field <- getNodeSet(XML.parsed, paste("//", field, sep=""))
  results <- t(sapply(parse.field, function(x) xmlAttrs(x)))
  if(typeof(results)=="list"){
    do.call(rbind.fill, lapply(lapply(results, t), data.frame
      , stringsAsFactors=F))
  } else {
    as.data.frame(results, stringsAsFactors=F)
  }
}
```

We illustrate the use of the `grabXML` function on the `<pitch>` nodes. This function requires a parsed XML document as the first argument (the previously created `xmlGame` is used here) and a character string indicating the nodes on which the process is applied (`"pitch"` here).

```
pitchesData <- grabXML(xmlGame, "pitch")

dim(pitchesData)
[1] 279   39
```

For the June 13, 2012 game between Houston and San Francisco, data on 279 pitches are parsed into the `pitchesData` data frame which consists of 279 rows and 39 columns. Functions specifically devised for retrieving PITCHf/x data are available in the `pitchRx` package, introduced in Chapter 12.

B.4   Details on the Data

In this section we describe the `atbat` and `pitch` attributes most frequently used in baseball analysis. Good collections of descriptions of the PITCHf/x attributes are provided by Mike Fast at `fastballs.wordpress.com/2007/08/02/glossary-of-the-gameday-pitch-fields/` and Alan Nathan at `webusers.npl.illinois.edu/~a-nathan/pob/tracking.htm`, and our descriptions that follow heavily draw from those sources.

B.4.1   atbat attributes

- **batter** and **pitcher**: MLBAM six-digit codes identifying the batter and the pitcher.

- **stand** and **p_throws**: handedness of the batter and the pitcher for the at-bat.

- **des** and **event**: detailed and short descriptions of the at-bat outcome.

B.4.2   pitch **attributes**

- **des**: Short description of the pitch outcome.

- **type**: One-character code of the pitch outcome (one among B=ball, S=strike, X=ball in play).

- **x** and **y**: Horizontal and vertical coordinates of the pitch as it crosses home plate, manually recorded. *Note: these are not the coordinates recorded by the PITCHf/x system.*

- **sv_id**: Timestamp denoting the second when PITCHf/x detected the pitched ball, formatted as YYMMDD_hhmmss.

- **start_speed** and **end_speed**: Speed in miles per hour at the release point and when the ball crosses the front of home plate.

- **sz_top** and **sz_bot**: Vertical coordinates for the top and the bottom of the strike zone of the batter currently at the plate. Both variables are expressed as feet from the ground and they are manually recorded at the beginning of every at-bat.

- **pfx_x** and **pfx_z**: Horizontal and vertical movement of the pitch compared to a theoretical pitch of the same speed with no spin-induced movement. Both variables are measured in inches.

- **px** and **pz**: Horizontal and vertical location of the pitch, measured when the pitch crosses the front of home plate as recorded by the PITCHf/x system. The coordinate system is centered on the middle of home plate and at ground level and viewed from the catcher/umpire point of view, thus a positive value of **px** indicates the pitch crosses the plate to the right of its middle and a negative value to the left. A negative value of **pz** indicates a pitch that bounced before reaching home plate. Both **px** and **pz** variables are measured in feet.

- **x0, y0, z0**: Coordinates indicating the calculated position of the ball at the release point. The **y0** parameter indicates the distance from home plate and is generally set at 50 feet from home plate[3]; researchers have found 55 feet as a distance that better approximates the true release point of the pitch and it is thus advisable to recalculate the coordinates at the 55ft mark,[4] as illustrated in Section B.6.1. **x0**, **y0** and **z0** are the left and

---

[3]In the first years of PITCHf/x this value varied from 40 to 55 feet.

[4]PITCHf/x data provided in the data frames of this book have undergone such transformation.

right position and the height of the release point in the same coordinate system as px and pz.

- **vx0**, **vy0**, and **vz0**: Components of the pitch velocity in three dimensions, measured at release in feet per second.

- **ax**, **ay**, and **az**: Components of the pitch acceleration in three dimensions, measured at release in ft/s^2.

- **break_y**, **break_angle**, and **pitch_type**: Quantities defining the "break" of a pitch by comparing its trajectory to that of a straight line going from release point to location at home plate. The variable **break_length** is the maximum deviation from the straight line in inches, **break_y** is the distance from home plate where such deviation occurs (in inches), and **break_angle** is the direction of the deviation, with positive values indicating a break toward a right-handed hitter.[5]

- **pitch_type**: Two-character abbreviation of the type of pitch as classified by the MLBAM algorithm. See Table B.4.2 for the decoding of the abbreviations.

- **spin_dir**: Direction of the spin of the ball, where 0° indicates a perfect top spin and 180° indicates a perfect bottom spin.

- **spin_rate**: Spin rate of the ball in revolutions per minute.

B.4.3 hip attributes (hit locations data)

While the attributes of both the atbat and pitch elements are located in the inning_all.xml file, hit location data are stored in the inning_hit.xml file, and the elements whose attributes are to be parsed are named hip.

- **des**: Short description of the batted ball outcome.

- **x** and **y**: Horizontal and vertical coordinates of the batted ball, manually recorded when the ball is first touched by a fielder (at the landing spot for home runs). The Gameday stringer marks the spot on a 250×250 pixel diagram of the ballpark; the position of home plate and the pixels-to-feet ratio are different from park to park.

---

[5]For a visual explanation of the three *break* attributes, see the figure displayed at the end of Mike Fast's glossary.

TABLE B.1
Key for the pitch types abbreviations used by MLBAM.

abbreviation	description
AB	Automatic Ball
CH	Change-up
CU	Curveball
EP	Eephus pitch
FA	Fastball (unspecified)
FC	Cut-fastball (cutter)
FF	Four-seam fastball
FO	Forkball
FS	Split-fingered fastball (splitter)
FT	Two-seam fastball
IN	Intentional ball
KC	Knuckle curve
KN	Knuckleball
PO	Pitchout
SC	Screwball
SI	Sinker
SL	Slider
UN	Unknown pitch type

B.5 Special Notes About the Gameday and PITCHf/x Data

Gameday and PITCHf/x have provided baseball analysts with an incredible wealth of data, but such amount of information should be used keeping in mind some issues that are tied to it, some of which are illustrated in Mike Fast's article *The Internet cried a little when you wrote that on it.*[6]

- **Release point estimate**: the release point is not directly tracked by PITCHf/x, but inferred by calculating the full trajectory of the pitch. Pitchers actually release the ball at different distances from home plate. That fact combined with the different trajectory of pitch types adds systematic errors in the release point values which are estimated a fixed distance from home plate. This issue should make analysts very careful when presenting evidence of release point changes for a particular pitcher, especially if comparisons come from different ballparks.

- **Pitch classification**: The pitch type associated to each tracked pitch (see Table B.4.2) is obtained by an algorithm developed by MLBAM which

---

[6]www.hardballtimes.com/main/article/the-internet-cried-a-little-when-you-wrote-that-on-it/.

makes use of the PITCHf/x data. The algorithm has been altered through the years. While the modifications have generally improved the accuracy of the classification of pitches, the year-to-year comparison of pitchers' repertoires require extra caution, as differences might be a product of the changes in the classifying algorithm.

• **Batted ball locations**: Batted ball location data are recorded manually and suffer from several problems. First, coordinates systems vary from ballpark to ballpark, as stringers mark spots on 250×250 pixel field diagrams with inconsistent home plate positioning and pixel-to-feet ratio. Second, researchers have shown that biases exist due to both the position (i.e. height) the stringer is assigned at the ballpark and the outcome of the batted ball. Finally, the stringers are instructed to mark the place where the ball is collected by a fielder and in case of deflections or caroms off the walls, it is impossible to infer the original angle of the batted ball.

B.6 Miscellanea

B.6.1 Calculating the pitch trajectory

As seen in the previous sections, PITCHf/x tracks data on location, velocity, and acceleration of a pitch. Using the kinematics equation for constant acceleration, the position of the ball at a given time t can be determined by the following equations:

$$x = x_0 + xv_0t + \frac{1}{2}axt \tag{B.1}$$

$$y = y_0 + yv_0t + \frac{1}{2}ayt \tag{B.2}$$

$$z = z_0 + zv_0t + \frac{1}{2}azt \tag{B.3}$$

The previous equations are translated to R with use of the following function `pitchloc`.[7]

```
pitchloc <- function(t, x0, ax, vx0, y0, ay, vy0, z0, az, vz0) {
  x <- x0 + vx0 * t + 0.5 * ax * I(t ^ 2)
  y <- y0 + vy0 * t + 0.5 * ay * I(t ^ 2)
  z <- z0 + vz0 * t + 0.5 * az * I(t ^ 2)
  if(length(t) == 1) {
    loc<-c(x, y, z)
```

---

[7]The code in this section has been slightly adapted from `code.google.com/p/r-pitchfx/`.

```
  } else {
    loc <- cbind(x, y, z)
  }
  return(loc)
}
```

The function `pitch.trajectory` calculates the trajectory of a pitch from release point to home plate at specified time intervals (the default choice of the argument `interval` is 0.01 seconds).

```
pitch.trajectory <- function(x0, ax, vx0, y0, ay, vy0, z0, az, vz0,
                        interval = .01) {
  cross.plate <- (-1 * vy0 - sqrt(I(vy0 ^ 2) - 2 * y0 * ay)) / ay
  tracking <- t(sapply(seq(0, cross.plate, interval), pitchloc, x0 = x0,
    ax = ax, vx0 = vx0, y0 = y0, ay = ay, vy0 = vy0, z0 = z0, az = az,
    vz0 = vz0))
  colnames(tracking) <- c("x", "y", "z")
  tracking <- data.frame(tracking)
  return(tracking)
}
```

B.6.2   An R package for getting and visualizing PITCHf/x data: `pitchRx`

The `pitchRx` package, contributed by Carson Sievert, provides functions for obtaining PITCHf/x data from MLBAM and for producing advanced visualizations from this data. The `scrapeFX` is a convenient function for downloading the data from MLBAM and storing them in R objects; by specifying a starting and an ending date, one obtains PITCHf/x data for every game played between the given dates.[8]

The following code uses `scrapeFX` to download PITCHf/x data for the games played between May 31, 2012 and June 1, 2012, and save the results in the object `dat`. This object consists of a list of two data frames, one containing information at the plate appearance level (`atbat`), and the second at the pitch-by-pitch detail (`pitch`) including pitch speed, movement, location, and type.

```
library(pitchRx)
dat <- scrapeFX(start="2012-05-31", end="2012-06-01")
```

The `pitchRx` package contains an example dataset `pitches`, containing data on every four-seamer and cutter thrown by either Mariano Rivera or Phil Hughes during the 2011 season. The `strikeFX` function in this package is used for creating a heatmap of the pitch locations, split by pitcher and opponent's handedness; the result is displayed in Figure B.3.

---

[8]Downloading data requires a few seconds per game, thus massive downloading of data requires plenty of time.

```
strikeFX(pitches, geom="tile"
  , layer=facet_grid(pitcher_name ~ stand))
```

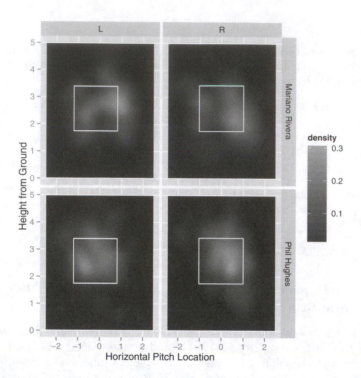

FIGURE B.3
Locations of four-seamers and cutters delivered by Mariano Rivera and Phil Hughes in 2011, by batter handedness.

The `animateFX` function produces several plots that, if displayed sequentially, create animations of the pitch trajectories from the pitcher's release point to home plate. With the following code one obtains an animation of Rivera's and Hughes' pitches, which is not displayed.

```
animateFX(pitches, layer=facet_grid(pitcher_name ~ stand))
```

The reader is encouraged to read *Carson's Personal Blog*[9] and particularly the pitchRx demo page[10] to fully appreciate the potential of the `pitchRx` package.

---

[9]cpsievert.wordpress.com/
[10]cpsievert.github.io/pitchRx/demo/

B.6.3 Cross-referencing with other data sources

The `master` table in Lahman's database is a useful resource for cross-referencing players across several data sources such as the *Baseball-Reference* website and the *Retrosheet* files. Unfortunately it currently does not contain a column for the MLBAM player identifier, thus the `master` table is not useful for merging PITCHf/x data to information coming from other sources. Baseball Prospectus currently provides a comprehensive and up-to-date list of players featuring both the Retrosheet and the MLBAM identifiers at `www.baseballprospectus.com/sortable/playerid_list.php`. The csv file linked at the top of the page can be read directly into R using the `read.csv` function.

```
players <- read.csv("www.baseballprospectus.com/sortable/
  playerids/playerid_list.csv")
```

Recently another very comprehensive source of players' identifiers has been made available online: The Register at Chadwick Baseball Bureau. At `chadwick-bureau.com/the-register/` one finds a link for the download of a zip file containing a register of players, managers, and umpires at any professional level (including, other than the Major Leagues, the Minor and Independent Leagues, Winter Leagues, Japanese and Korean top levels, and the Negro Leagues).

B.6.4 Online resources

The following PITCHf/x resources are currently available on the World Wide Web. Note that, due to site maintainers being hired by MLB front offices or exclusive licensing contracts, these resources are subject to being removed or moved.

- **Brooks Baseball** (`www.brooksbaseball.net/`): Created and maintained by Dan Brooks, its main features are the *Player Cards*, consisting of tables and charts for every pitcher who has ever played in a ballpark with the PITCHf/x system installed. Tables and charts report information on characteristics of pitches, their usage (including sequencing), and the outcomes they produce. The classification of pitches used at Brooks Baseball is not the MLBAM one, as pitches are classified by *Pitch Info LLC*. Another useful resource of Brooks Baseball is the *PitchFX Tool*, which allows site visitors to select one pitcher for one game and obtain a pitch-by-pitch table.

- **Baseball Prospectus** (`www.baseballprospectus.com/`): In its *Statistics* section, Baseball Prospectus offers *PITCHf/x Hitters Profiles*, *PITCHf/x Pitchers Profiles*, *PITCHf/x Leaderboards*, and *PITCHf/x Matchups*. The building blocks of these resources come from the previously mentioned Brooks Baseball.

- **FanGraphs** (`www.fangraphs.com/`): FanGraphs has PITCHf/x tables and charts for individual players. For example, pitcher James Shields's PITCHf/x page is available at `www.fangraphs.com/pitchfx.aspx?playerid=7059&position=P`.

- **F/X by Texas Leaguers** (`pitchfx.texasleaguers.com/`): Allows one to set a time frame and find PITCHf/x pitching or batting data for one particular player. This site includes charts on trajectory and movement, tables on pitch characteristics, and outcomes and pitcher/batter match-ups.

- **Prof. Alan Nathan's The Physics of Baseball** (`webusers.npl.illinois.edu/~a-nathan/pob/index.html`): Contains research on baseball physics and has a section dedicated to pitch tracking using video technology at `webusers.npl.illinois.edu/~a-nathan/pob/pitchtracker.html`.

- **Katron's MLB Gameday BIP Location** (`katron.org/projects/baseball/hit-location/`): Allows to transpose hit location data of a given ballpark in another ballpark of choice. Keeping in mind all the caveats previously illustrated for batted ball data, it can be used to explore the effect moving to a new team can have on a player's batting.

- **Sportvision** (`www.sportvision.com/baseball`): Sportvision is the company which has devised the PITCHf/x system. On its website, videos for other products they exclusively license to MLB teams (such as HITf/x, FIELDf/x and COMMANDf/x) are featured.

Bibliography

[1] Adler, J. (2006), *Baseball Hacks: Tips & Tools for Analyzing and Winning with Statistics*, O'Reilly Media.

[2] Albert, J. (2002), "Smoothing career trajectories of baseball hitters," manuscript, Bowling Green State University.

[3] Albert, J. and Bennett, J. (2003), *Curve Ball: Baseball, Statistics, and the Role of Chance in the Game*, Springer, New York.

[4] Albert, J. (2003), *Teaching Statistics Using Baseball*, Mathematical Association of America.

[5] Albert, J. (2008), "Streaky hitting in baseball," *Journal of Quantitative Analysis in Sports*, 4, 1.

[6] Albert, J. (2009), "Is Roger Clemens whip trajectory unusual?" *Chance*, 22, 2, 9-20.

[7] Albert, J. and Rizzo, M. (2012), *R by Example*, Springer, New York.

[8] Allen, D. (2009a), "Deconstructing the Non-Fastball Run Maps", *Baseball Analysts* website, http://baseballanalysts.com/archives/2009/03/deconstructing_1.php.

[9] Allen, D. (2009b), "Platoon Splits for Three Types of Fastballs", *Baseball Analysts* website, http://baseballanalysts.com/archives/2009/05/platoon_splits.php.

[10] Berry, S., Reese, S., and Larkey, P. (1999), "Bridging different eras in sports," *Journal of the American Statistical Association*, 94, 661-676.

[11] Berry, S. (1990), "The summer of '41: a probabilistic analysis of DiMaggio's streak and Williams's average of .406," *Chance* 4, 4, 8-11.

[12] Bradley, R. and Terry, M. (1952), "Rank analysis of incomplete block designs: I. The method of paired comparisons," *Biometrika*, 39, 324-345.

[13] Bukiet, B., Elliotte R., and Palacios, J. (1997), "A Markov chain approach to baseball," *Operations Research*, 45, 14-23.

[14] Cleveland, W. (1979), "Robust locally weighted regression and smoothing scatterplots," *Journal of the American Statistical Association*, 74, 829–83.

[15] Cleveland, W. (1994), *Elements of Graphing Data*, Hobart Press.

[16] Caola, R. (2003), "Using calculus to relate runs to wins: Part i," *By the Numbers*, 13, 9–16.

[17] Davenport, C. and Woolner K. (1999), "Revisiting the Pythagorean Thorem," *Baseball Prospectus* website, `www.baseballprospectus.com/article.php?articleid=342`.

[18] Dewan, J. (2006), *The Fielding Bible*. ACTA Publications.

[19] Dewan, J. (2009), *The Fielding Bible Volume II*. ACTA Publications.

[20] Dewan, J. and Jedlovec, B. (2012), *The Fielding Bible Volume III*. ACTA Publications.

[21] Dolphin, A., Lichtman, M. and Tango, T. (2007), *The Book: Playing the Percentages in Baseball*, Potomac Books Inc.

[22] Fair, R. (2008), "Estimated Age Effects in Baseball," *Journal of Quantitative Analysis of Sports*, 4, 1.

[23] Fast, M. (2009), "What the heck is pitchf/x?" In *The Hardball Times Baseball Annual 2010*. ACTA Publications.

[24] Gould, S. (1989), "The streak of streaks," *Chance*, 2, 2, 10-16.

[25] Heipp, B. (2003), "W% estimators," *Buckeyes and Sabermetrics* website, `gosu02.tripod.com/id69.html`.

[26] Humphreys, M. (2011), *Wizardry: Baseball's All-Time Greatest Fielders Revealed*. Oxford University Press.

[27] James, B. (1980), *Bill James Baseball Abstract*, self-published, Lawrence, KS.

[28] James, B. (1982), *Bill James Baseball Abstract*, Ballantine Books.

[29] Kabacoff, R. (2011), *R in Action*, Manning Publications.

[30] Kemeny, J. and Snell, L. (1976), *Finite Markov Chains*, Vol. 210, Springer-Verlag, New York.

[31] Keri, J. and Baseball Prospectus (2007), *Baseball Between the Numbers: Why Everything You Know about the Game Is Wrong*, Basic Books.

[32] Lahman, S. (2012), "Lahman's Baseball Database, 1871-2012, v. 2012, Comma-delimited version," `seanlahman.com/files/database/lahman2012-csv.zip`.

[33] Lindsey, G. (1963), "An investigation of strategies in baseball," *Operations Research*, 11, 4, 477-501.

[34] Marchi, M. (2010), "Platoon splits 2.0", *Hardball Times* website, `http://www.hardballtimes.com/main/article/platoon-splits-2.0/`.

[35] McCotter, T. (2009), "Hitting streaks don't obey your rules: Evidence that hitting streaks aren't just by-products of random variations," *The Baseball Research Journal*, 37, 62-70.

[36] Murdoch, D.J. and Chow, E.D. (1996), "A graphical display of large correlation matrices," *The American Statistician*, 50, 178-180.

[37] Murrell, P. (2006), *R Graphics*, Chapman and Hall, Boca Raton, Florida.

[38] Nathan, Alan M. (2013), The *Physics of Baseball* website, `baseball.physics.illinois.edu`.

[39] Palmer P. (1983), "Balls and Strikes," In *Baseball Analyst*, Issue 5, February 1983. Available at `http://sabr.org/research/baseball-analyst-archives`.

[40] Pankin, M. (1987), "Baseball as a Markov chain," In *The Great American Baseball Stat Book*, 520-524.

[41] R Development Core Team (2013), "R: A language and environment for statistical computing," R Foundation for Statistical Computing, Vienna, Austria, `www.R-project.org`.

[42] Rickey B. (1954), "Goodby to some old baseball ideas," In *LIFE*, August 2, 1954 issue. Available at `goo.gl/mZiG5`.

[43] RStudio (2013). RStudio: Integrated development environment for R (Version 0.97.336) [Computer software]. Boston, MA. Retrieved June, 22, 2013. Available from `www.rstudio.org`.

[44] Sarkar, D. (2008), *Lattice: Multivariate Data Visualization with R (Use R!)*, Springer, New York.

[45] Schwarz, A. (2005), *The Numbers Game: Baseball's Lifelong Fascination with Statistics*, St. Martin's Griffin.

[46] Seidel, M. (2002), *Streak: Joe DiMaggio and the summer of '41*, Bison Books.

[47] Star, J. (2011), "The Road to October: Sept. 29, 2011," *MLB* website, http://mlb.mlb.com/news/article.jsp?content_id= 25380714&vkey=roadtooctober2011&ymd=20110929.

[48] Triumph Books (2012), *2012 Official Rules of Major League Baseball*, Triumph Books.

[49] Venables, W. N., Smith, D. M., and the R Development Core Team (2011). "An Introduction to R," Version 2.13.0 (2011-04-13).

[50] Walsh, J. (2008), "Searching for the games best pitch," *Hardball Times* website, http://www.hardballtimes.com/main/article/ searching-for-the-games-best-pitch/.

[51] Walsh, J. (2010), "The Compassionate Umpire," *Hardball Times* website, http://www.hardballtimes.com/main/article/ the-compassionate-umpire/.

[52] Wickham, H. (2009), *ggplot2: Elegant Graphics for Data Analysis*, Springer, New York.

[53] Wilkinson, L. (2005), *The Grammar of Graphics*, second edition, Springer, New York.

Index

D, 100
SMA, 240
abline, 82, 91, 192
abs, 93
aggregate, 97, 107, 112, 135, 136, 245, 251
animateFX, 322
annotation_raster, 157
as.character, 108
as.data.frame, 97
as.integer, 135
as.numeric, 108
axis, 165
barplot, 38, 60
boxplot, 74
by, 41, 48
cbind, 45
coef, 91
color2D.matplot, 164
colwise, 291
contourLines, 181
contourplot, 176
coord_equal, 148
cor.plot, 294
cumsum, 35, 78
curve, 71
cut, 60
c, 32
data.frame, 51, 194
dbConnect, 262
dbDisconnect, 263
dbWriteTable, 266
ddply, 52, 109, 193, 194
densityplot, 132
dev.off, 63
dimnames, 38, 109
direct.label, 297

dotchart, 54, 64
dotplot, 136
droplevels, 136
expand.grid, 101, 163, 174
expression, 100
facet_wrap, 149, 198
factor, 39, 48
format, 135
geom_line, 152, 198
geom_path, 151
geom_point, 146
geom_segment, 151
geom_vline, 152
getNodeSet, 315
ggplot, 146
grabXML, 316
grepl, 167
grep, 167
gsub, 167
guides, 151
head, 47
histogram, 132
hist, 66
identify, 68, 98
ifelse, 77, 225
lattice package, 129
ldply, 316
legend, 79, 82
length, 34
lines, 67, 79, 82, 115, 192
list, 40
lm, 89, 191
loadWorkbook, 286
loess, 174
log, 95
lowess, 53, 67, 115
mapply, 170

`mat.sort`, 294

`matrix`, 37, 109

`max`, 34

`mean`, 34

`melt`, 296

`merge`, 49, 52, 114

`mtext`, 165

`names`, 80

`nchar`, 165

`ncol`, 107

`nrow`, 93

`odbcClose`, 264

`odbcConnect`, 264

`order`, 34, 47, 51

`par`, 74

`parse.retrosheet.pbp`, 301

`paste`, 106

`pie`, 62

`plotcorr`, 292

`plot`, 33, 43, 53, 79, 82

`png`, 63

`points`, 91

`predict`, 192

`prop.table`, 119, 179, 215

`rbind`, 49

`rbinom`, 225

`read.csv`, 44

`readJPG`, 156

`readWorksheet`, 286

`recode`, 188

`replicate`, 217, 247

`rmultinom`, 226

`round`, 42, 119

`rug`, 240

`sample`, 130

`sapply`, 51, 116, 194, 245

`scale_linetype_discrete`, 272

`scrapeFX`, 321

`sd`, 34

`seq`, 33, 101

`setdiff`, 289

`shell`, 273

`sort`, 34

`sqlFetch`, 268

`sqlQuery`, 269

`sqlSave`, 265

`sqrt`, 93

`stat_binhex`, 155

`stats_summary`, 272

`stringdistmatrix`, 289

`strikeFX`, 321

`stripchart`, 66, 74, 112

`str`, 39

`subset`, 48, 49, 51, 52

`substr`, 242

`summaryBy`, 287

`summary`, 35, 46, 192

`sum`, 34

`table`, 38, 39, 60, 119, 179

`tail`, 88

`text`, 72, 82, 91

`truehist`, 120

`unique`, 51

`windows`, 66

`write.csv`, 45

`wtd.cor`, 292

`xmlAttrs`, 315

`xmlParse`, 314

`xtabs`, 101

`xyplot`, 135, 173, 269

Aaron, Hank, 2, 6, 52, 53, 77

Adler, Joseph, 314

age, of MLB player, 77

Alomar, Roberto, 199

artificial turf, 281

Astrodome, 279

axes labels, 61

Aybar, Erick, 246

bar graph, 38, 60

Baseball Heat Maps website, 274, 314

Baseball Prospectus website, 107, 323

Baseball Reference website, 162, 238, 323

Baumer, Ben, 126

Bautista, Jose, 21, 116

Berkman, Lance, 257

Berry-Terry model, 223

Bichette, Dante, 278

Biggio, Craig, 199, 280
birth year, 188
Bonds, Barry, 2, 77
bootstrap procedure, 272
boxplots
    parallel, 73
bracket notation
    for data frames, 46
    for lists, 40
    for vectors, 35
Brock, Lou, 6
Brooks, Dan, 323
Burks, Ellis, 278

Cabrera, Miguel, 116, 145, 173
car package, 188
Cashman Field, 277
Castilla, Vinnie, 278
character variable, 37
clumpiness
    measure of, 246
Colorado Rockies, 271
comma separated value format, 44
conditional density plot, 133
connPackage package, 267
console window, 32
containers in R, 37
contour plot, 174
Coors Field, 271, 276

Damon, Johnny, 199
Davis, Eric, 206
defensive runs saved, 285
density plot, 132
DiMaggio, Joe, 238
directlabels, 297
DiSarcina, Gary, 206
Division series
    American League, 19
doBy package, 287
Dodgers Stadium, 277
dot plot, 63

Eisenreich, Jim, 206
ellipse package, 292
Ellis, Mark, 246

eras of baseball, 60
Espinoza, Alvaro, 206
extreme values
    in residual graph, 92

facets
    in ggplot2 package, 149
factor variable, 38
FanGraphs website, 288
fans scouting report, 285
Fenway Park, 276
Fielder, Prince, 87, 116
FIP measure, 47
Flip Play, 20
Foxx, Jimmie, 65, 72
Franco, Julio, 199
frequency table, 38, 60, 179
Frisch, Frankie, 199
function in R, 42
FX by Texas Leaguers, 324

Gardner, Brett, 87
Gehrig, Lou, 16, 68, 72
ggplot2 package, 129, 198, 255, 271, 296
Giambi, Jeremy, 19
glyph graph, 292
Green Monster, 276
Greenberg, Hank, 72

Hall of Fame, 59
Hardball Times website, 99
heat map, 163
histogram, 66, 132
hitting streak, 238
Hornsby, Roger, 72
Hughes, Phil, 321
Humber, Phil, 22

Ibanez, Raul, 246, 249
installing package, 50

Jeter, Derek, 19, 199, 294
jittering method, 66
jpeg package, 156

Katron's MLB gameday BIP location, 324
Kemp, Matt, 116
Killebrew, Harmon, 52

Lahman
    database, 4, 50, 77, 273
    package, 50
Lahman, Sean, 3
lattice package, 173, 269
linear regression, 89
list variable, 40
loading package, 50
loess smoother, 67, 115, 174, 203
logical operators, 35
logical variable, 35
logistic model, 231
Long, Terrence, 19

Mantle, Mickey, 14, 43, 188
Maris, Roger, 14
Markov chain, 212
MASS package, 120
Mathews, Eddie, 197, 199
Matthews, Gregory, 126
matrix variable, 37
Mays, Willie, 52
McGehee, Casey, 246
McGwire, Mark, 14, 79
MLBAM Gameday application, 21
moving average, 240
moving average plot, 243
multipanel conditional display, 133, 149, 198
Mussina, Mike, 19, 162

Napoli, Mike, 116
Nathan, Alan, 23, 274, 316, 324
New York Yankees, 13
normal distribution, 225

Oakland-Alameda County Coliseum, 277
objects in R, 36
Olivo, Miguel, 246
on-base percentage, 69

openWAR package, 126
OPS measure, 67
outliers, 75

package TTR, 240
panel function
    in lattice graphics, 138
park factors, 275
peak age, 191
perfect game, 22
Phillips, Tony, 206
pie chart, 62
pitch
    movement, 23
    speed, 23
    trajectory, 23
PITCHf/x data, 274
pitchRx package, 321
plotrix package, 164
plyr package, 52, 291
Posada, Jorge, 19
post season
    baseball, 226
predict
    linear model, 91
psych package, 294
Pujols, Albert, 111
Pythagorean expectation, 93

quadratic fit, 190

random model, 246
reshape2 package, 296
residual graph, 91
residuals
    from linear model, 91
Retrosheet, 264
revised zone rating, 285
Ripken, Cal, 16
Rivera, Mariano, 321
RMySQL package, 262
Robinson, Frank, 52
RODBC package, 263
Rodriguez, Alex, 77
Rollins, Jimmy, 294
root mean square error, 93

RStudio, 30
Ruiz, Carlos, 246
run differential, 89
runs
    in remainder of inning, 106
    value, 110
runs expectancy, 105
runs potential, 217
Ruth, Babe, 8, 11, 65, 68, 72, 75, 77
Ryan, Nolan, 280

saving graph
    using R functions, 63
    using RStudio, 62
scatterplot, 33, 53, 67, 115, 135, 146,
        173
schedule
    of baseball games, 224
Schmidt, Mike, 197, 199
script of expressions, 41
Sheffield, Gary, 197
Sievert, Carson, 321
similarity scores, 195
simulation experiment, 246
slugging percentage, 69
Smith, David, 14
smoothing count, 222
smoothing curve, 53
Sosa, Sammy, 14, 79, 197
sourcing a file, 41
Spahn, Warren, 31
Spencer, Shane, 19
Sportvision, 21, 324
spray chart, 146
standard deviation, 252
string variable, 37
stringdist package, 289
stripcharts
    single, 66, 112, 206
    parallel, 73
superposed display, 134
Suzuki, Ichiro, 242

Thomas, Frank, 197, 199
total zone rating, 285

transition probabilities, 212

ultimate zone rating, 285
umpires, 173
Upton, Justin, 253

vector structure, 32
Verlander, Justin, 130, 173, 179
vertical line graph, 62
Votto, Joey, 116

Walker, Larry, 278
weighted on-base percentage, 249
weights package, 292
Wilkomm, Kyle, 314
Williams, Ted, 65, 72, 257
winning percentage, 89
working directory, 30
workspace
    in R, 130
World Series, 37

XAMPP distribution, 260
XLConnect package, 286
XML package, 314

Zimmerman, Jeff, 314